T0093787

White Elephant Technology

White Elephant Technology

50 Crazy Inventions That Should Never Have Been Built, And What We Can Learn From Them

JOHN J. GEOGHEGAN
PHOTO EDITOR ERIC MILES

For anyone who has ever tried doing something different and was told not to.

Cover illustrations: Front: (top) the AVE Mizar flying car; (bottom) the New York Central's jet-powered M- 497 railcar. *Back:* the Office Isolator.

First published 2023

The History Press
97 St George's Place, Cheltenham,
Gloucestershire, GL50 3QB
www.thehistorypress.co.uk

British Library Cataloguing in Publication Data.
A catalogue record for this book is available from the British Library.

ISBN 978 1 80399 014 9

Typesetting and origination by The History Press
Printed in Turkey by IMAK

Let us be thankful for the fools. But for them the rest of us could not succeed.

Mark Twain

Contents

Introduction

> We are such stuff as dreams are made on.
> William Shakespeare, *The Tempest*, Act IV, Scene 1

What exactly is White Elephant Technology?

Simply put, any unusual invention past or present that fails to find a market despite its innovative nature qualifies as White Elephant Technology, or WETech for short.

You're already familiar with the usual suspects such as flying cars and jet packs, but the majority of inventions covered in this book are ones most people have never heard of. From jeeps that fly to tanks that shouldn't; from a wave-powered boat that takes forever to reach its destination to a jet-powered train that shook itself apart, *White Elephant Technology* showcases each inventor's talent for inventing something nobody asked for. Importantly, none of these inventions are speculative. Each one was built, field tested, and worked more or less as planned except in the case when it inadvertently killed its inventor.

So, why is the study of WETech inventions important?

It's not always easy to understand why one invention succeeds and another fails. Luck, timing and market conditions have as much to do with it as functionality. Put another way, success for even the cleverest inventor sometimes comes down to a roll of the dice.

When Dean Kamen introduced the Segway in 2001, it was predicted his invention would so revolutionise personal transportation as to become ubiquitous. Kamen even claimed his invention would be the fastest in history to reach $1 billion in sales. And yet, the Razor, a foot-powered scooter introduced the year before for $29.99, sold 50 million units while the Segway sold only 140,000. Despite tremendous hype and financial backing, the Segway was a failure.

Even though we're told there's more to be learned from failure than success, there's no denying our culture prizes success more. Yet this might change if more people realised just how often commercial endeavours fail. For example:

- More than 80 per cent of all books, movies, popular music and video games fail to turn a profit
- 85 per cent of all new grocery products are pulled from supermarket shelves within a year of introduction
- And 90 per cent of all patented inventions never earn a dime*

In other words, failure is the rule not the exception.

And high failure rates aren't confined just to business. Take sport, for example. A baseball batting average of .300 is considered excellent, even though it means a batter fails to hit the ball every seven out of ten times they bat.

Put simply, failure is a significant part of life. That's why the First Rule of Failure states: the majority of commercial endeavours are *unlikely* to succeed. This alone makes WETech inventions worthy of study.

High failure rates don't mean we should stop striving, however. After all, where would we be if Columbus hadn't failed at finding a better trade route to India, or we'd given up on reaching the Moon after the Apollo 1 fire? The world would be a lesser place. Simply put, we're better off for trying even when we fail.

Obviously, failure has a lot to teach us. What's often overlooked, however, is the tremendous amount of talent, perseverance and sheer out-of-the-box thinking that goes into creating something even when it fails. That's why the ingredients necessary for

* A total of 600,000 patent applications are submitted in the United States each year, of which 326,000 are approved. Source: www.patentrebel.com. Only 2–10 per cent of approved patents earn enough money to maintain their protection. Source: www.inventiontherapy.com.

inventing something new (including grit, ingenuity and optimism) are the same ingredients for success.

Perhaps this is why research into failure is finally catching on. Once confined to academia, medicine and aviation, the study of failure now generates international conferences as well as books, magazine and newspaper articles and research studies. There's even a Museum of Failure that tours the world. Still, no one has ever undertaken a survey of failed inventions despite history being littered with them. Until now.

White Elephant Technology is written in a light-hearted, easy-to-understand manner. Each entry in the book's ten, thematically linked chapters explains who the inventor was, how their invention worked, why it failed and what, if anything, we can learn from their mistakes. Numerous photographs, diagrams and illustrations are included depicting what the invention (and in some cases, inventor) looked like. Naturally, humour is inevitable, but the overall take is respectful.

I've specialised in reporting on WETech inventions for twenty years. I've written about them for the *New York Times* Science section, *Popular Science* magazine and Smithsonian's *Air & Space* magazine among other publications. I've even written two non-fiction books and worked on two documentaries with WETech inventions at their core.

Although the technical aspect of many of these inventions is fascinating, what really drew my attention was the heroic investment every inventor makes in their invention, even when it fails. In other words, the story behind the invention is as important as the invention itself. Since many inventions represent years of hard work, financial sacrifice and fierce determination on the part of the inventor, White Elephant Technology is one of the purest expressions of the human condition. No wonder so many inventors refuse to give up.

Hopefully, *White Elephant Technology: 50 Crazy Inventions That Should Never Have Been Built, And What We Can Learn From Them* captures all the ambition, pitfalls and heartbreak that go into inventing. At the very least, it will cause you to shake your head in wonder while asking the question, 'What in the world were they thinking?' Such dedication on the part of an inventor may seem crazy when success remains elusive, but as an Apple computer commercial once noted, 'It's the people who are crazy enough to think they can change the world that do.'

Where would we be without them?

Chapter 1

Hybrids: Making a Boat a Plane

> I have been branded with folly and madness for attempting what the world calls impossibilities … but should this be all, I shall be satisfied.
>
> Richard Trevithick, nineteenth-century British inventor

Some WETech inventions take a mode of transportation and make it do something for which it was never intended. Boats by themselves do a great job of floating while planes comfortably navigate the sky. But combine the two and you compromise their original function, making the boat less seaworthy and the plane more likely to crash. These hybrid combos prove more isn't necessarily better and it's often less. Take, for example, the swimming tank.

1 The Swimming Tank

Tanks were known for two things during the Second World War: their mobile fire power and heavy armour, which shielded them against attack. What they weren't known for was swimming. But the Allies needed to get tanks ashore during D-Day to support the invasion. The goal, then, was to create an amphibious tank that could swim to the beachhead by itself rather than be carried by specialised landing craft. Given the M4A1 Sherman tank weighed 33 tons with armour 3in thick, you'd think making one float was a bad idea. However, such considerations have never stopped a WETech inventor.

Nicholas Straussler, a Hungarian engineer working for the British, was charged with finding the solution. Straussler was a specialist in designing amphibious, off-road vehicles for the military. Still, he faced a considerable problem – overcoming a tank's weight and lack of buoyancy. His answer was to develop a waterproof flotation device – a canvas skirt surrounding the outside of the tank that displaced enough water to enable it to float. The collapsible skirt, which left the tank's top and bottom open to the elements, was raised using compressed air. Once inflated, metal scaffolding was snapped into place to provide the skirt with additional support.

The D-Day invasion plan included offloading the swimming tanks 2 miles from shore. Since they sat low in the water, the tank part wasn't visible to the enemy. In fact, the surrounding skirt made it look like a boat. A periscope extending from the tank's turret enabled the driver to see where they were going, while they used a compass for navigation. Twin, three-bladed propellers underneath the tank's rear carriage delivered a maximum speed of 4 knots. They could also be swivelled left and right for steering. An automatic bilge kept the inside of the tank dry. As the tank approached shore, the front of its skirt could be collapsed like an accordion, allowing its 3in gun to fire. Once on land, the rest of the skirt was quickly deflated, enabling it to proceed as a conventional tank.

The DD M4A1 Sherman 'swimming' tank. (US Army)

The DD M4A1 Sherman tank (DD stood for Duplex Drive – its two means of propulsion) was a hybrid destined to failure. No one was surprised when military wags began calling it the 'Donald Duck' tank.

Straussler's invention faced several non-trivial problems. First, the tanks were so cumbersome they were difficult to steer in the ocean. Additionally, they sat so low in the water that anything higher than a 2ft wave risked swamping them. Unfortunately, the waves off Omaha Beach the morning of 6 June 1944 were 6ft high.

The moment of truth came when the five-man tank crews had to seal themselves inside their steel-plated coffin before being deployed 3 miles off shore in the middle of a storm. They must have known in the pit of their stomachs that things weren't going as planned, but they got in anyway – an amazing act of courage. It's a moment many WETech inventors experience, but in this case the swimming tank's inventor was safe, warm and dry while someone else paid for his mistakes.

A rear view of the DD M4A1 Sherman tank. (US Army)

Of the twenty-nine swimming tanks launched off Omaha Beach, twenty-seven sank like a stone, some with their tank crews trapped inside. Canadian and British tanks did somewhat better due to calmer seas, but of the 120 tanks launched that day at least forty-two (or more than a third) disappeared beneath the waves.

Straussler worked on a variety of projects after the war, most related to off-road or amphibious vehicles. He fell foul of the British authorities in 1957 when he was charged with violating export controls for selling a truck he'd modified to a Soviet-bloc country. In 1961, he sued the United States for patent infringement, claiming that amphibious military vehicles such as the 'Otter' and 'Duck' were based on his ideas. He lost the case.

Straussler, who had thirty patents to his name, continued working right up until his death in 1964 at age 75. Although he made other contributions to the war effort, he will always be remembered, not altogether fondly, as the father of the swimming tank.

Two Sherman Duplex Drive tanks recovered from the seabed after D-Day are preserved at the Musée des Epaves Sous-Marine du Débarquement, Port-en-Bessin-Huppain, Normandy, France. A swimming tank can also be seen at the Tank Museum in Dorset, England.

2 Underwater Aircraft Carriers

The idea of an underwater aircraft carrier – a giant submarine that could carry aeroplanes – may seem counter-intuitive but it made sense for certain seafaring countries in the days before radar.

Germany, Great Britain, the United States, Italy, France and Japan all experimented with sub–plane combinations, albeit with mixed results. As crazy as it might sound, there was a strategic reason for a submarine to carry an aeroplane. Subs in the first half

A captured Japanese I-400 submarine in Sasebo Bay.

of the twentieth century didn't just sink enemy ships, but were used as scouts to find the enemy fleet. They had a significant drawback, though. They rode so low in the water that their field of vision was limited to 7 miles. But you could dramatically improve a sub's scouting range if it carried, launched and retrieved its own aircraft. Hence, the underwater aircraft carrier was born.

Underwater aircraft carriers proved a White Elephant Technology for virtually all the countries that experimented with them. Nevertheless, Japan was determined to make plane-carrying subs a success. The island nation's experiments began in 1923 when it purchased a floatplane from Germany and began holding sea trials. A crane mounted on a submarine's deck lowered the seaplane over the side, where it could take off from the water.

As the Imperial Japanese Navy (IJN) grew in experience, it began adding watertight deck hangars to its subs for storing a collapsible biplane – an important step in integrating the seemingly incompatible. By the autumn of 1928, Japan had progressed far enough that the IJN was satisfied that a sub-plane combination was practical.

When the Second World War broke out, Japan was all in where underwater aircraft carriers were concerned. Eleven plane-carrying subs surrounded the island of Oahu during Japan's surprise attack on Pearl Harbor. Afterwards, one of the subs stayed behind to launch its plane to survey the damage. Eventually, Japan integrated plane-carrying subs into every combat theatre where its navy fought.

US Navy personnel inspecting the gun of an I-400.

Japan made innovative use of its underwater aircraft carriers. The *I-25** twice launched its floatplane in autumn 1942 to drop bombs on Oregon – the first time anyone used a submarine's aeroplane to attack an enemy's mainland. By war's end, Japan had built more than forty-one plane-carrying subs, making it the hands-down expert on this unusual weapon.

The culmination of the underwater aircraft carrier came in the form of Japan's I-400-class subs. The largest submarines ever built when they began commissioning in 1944, these behemoths were more than a football field long, had a crew

* Japan called its subs I-boats, 'I' being a transliteration of a Japanese character signifying the first (as well as the largest) class of submarines.

The last surviving Aichi M6A1 Seiran was flown by Imperial Japanese Navy Lt Kazuo Akatsuka from Fukuyama to Yokosuka, where he surrendered it to an American occupation contingent. (Transferred from the United States Navy, 1945)

of 200 men and could travel one and a half times around the world without refuelling. Additionally, each I-400-class sub carried three attack bombers in a watertight deck hangar, which it launched off its bow using a pneumatic catapult. Originally conceived by Admiral Isoroku Yamamoto as a follow-up punch

The watertight hanger of an I-400.

to his attack on Pearl Harbor, a squadron of I-400 subs was on its way to attack US naval forces when the war ended.

Despite Japan's success, underwater aircraft carriers never caught on as an offensive weapon, in part because technological advancements such as radar and sub-launched missiles made them unnecessary. Still, their legacy endures today in the form of 'boomers', ballistic missile subs capable of launching a nuclear attack against an enemy's mainland These are the true descendants of the I-400 class of subs, proving that some WETech inventions fail simply because they're ahead of their time.

The I-400 squadron's last surviving Aichi M6A1 attack bomber can been seen at the Smithsonian National Air and Space Museum's Steven F. Udvar-Hazy Center in Chantilly, Virginia.

3 Cromwell Dixon's Sky-Cycle

Cromwell Dixon's Sky-Cycle was unique for its time. Dixon took an everyday bicycle and converted it into a pedal-powered blimp. A true hybrid, Dixon's Sky-Cycle dispensed with the bicycle's two tyres while keeping its frame, handlebars, seat and pedals. Dixon even used the bicycle chain to drive a propeller. A silk envelope, 32ft long and 15ft wide, was filled with lighter-than-air hydrogen to keep the whole thing afloat. As Dixon pedalled, the bicycle chain drove a front-mounted propeller while the handlebars, connected by wires to a rear-mounted rudder, were used for steering. But the most novel aspect of Dixon's Sky-Cycle was the fact it was designed and built by a 14-year-old boy from Columbus, Ohio, in 1907.

The internal combustion engine was relatively new, heavy and unreliable, so not easily adapted to lighter-than-air flight. Besides, Cromwell's mother thought him too young to fly with something as dangerous as a motor. Dixon's ingenious solution was to use the proven alternative of pedal power.

To be clear, Dixon had help building his Sky-Cycle. His mother not only used her sewing machine to sew the craft's gas bag, she helped him raise money for the venture. She even permitted her son to build his Sky-Cycle in their back garden, which, given Dixon manufactured his own highly flammable hydrogen, must have alarmed the neighbours.

Dixon made his first test flight, wearing a cap and plus fours, from Driving Park in Columbus, Ohio, with an American flag trailing behind him.

The New York Times described him as propelling his Sky-Cycle through the air, 'easily and without trouble ... like some great bird'.

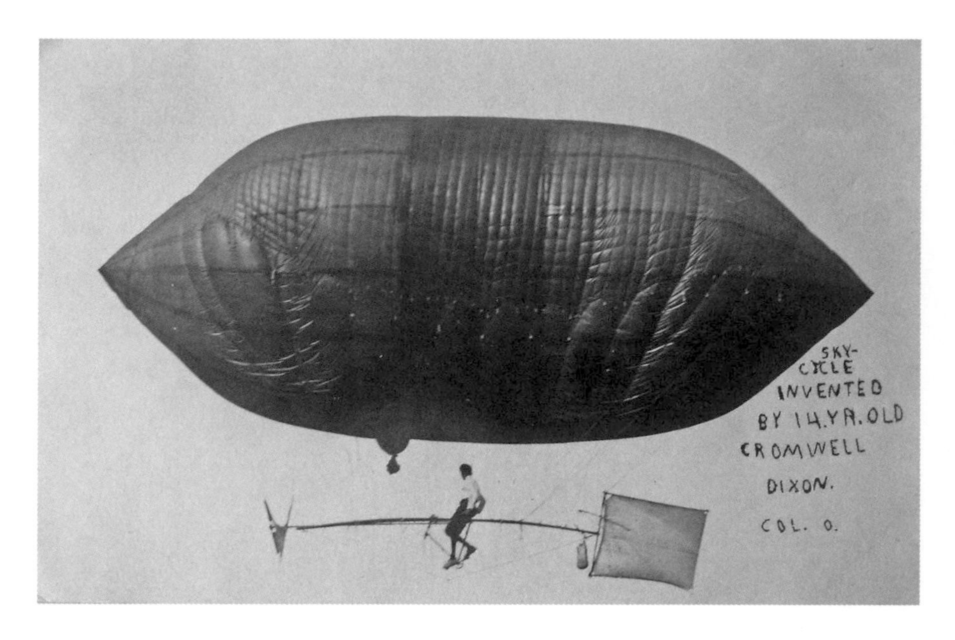

Cromwell Dixon steering the Sky-Cycle.

But Dixon's first flight did not go without a hitch. Although he reached an altitude of 2,500ft, the valve on his yellow-tinted gas bag came loose, allowing precious hydrogen to escape. As his airship began to descend, Dixon coolly slipped off his bike and, reaching up, sealed the valve before returning to his seat. But the Sky-Cycle had lost so much hydrogen it kept falling. Forced to throw whatever he could to lighten the weight, Dixon surrendered his coat and cap to the world below. Disaster was averted when he managed to land safely in a vacant parking lot.

Cromwell Dixon.

Despite the teething problems, *The Times* ranked Dixon, 'as one of the most successful aeronauts in the world'.

'Why ... it's easy,' Dixon told the paper of record. 'There's nothing to be afraid of.' His mother added, 'Cromwell is as happy as a King over his successful trip ... We are perfectly satisfied with the result.'

Like a lot of WETech inventors, Dixon had shown mechanical aptitude from an early age. Working in a barn behind his house, he built a metal swimming fish out of a wind-up clock. Fascinated by aeronauts of his day, he soon set his sights on becoming one of them.

Dixon held an exhibition flight one week after his first test flight, charging 25 cents admission to anyone who wanted to see him take to the sky. Unfortunately (and stories about WETech inventions are always filled with 'unfortunately'), a fire later destroyed Dixon's invention. Undeterred, he went to work on an improved Sky-Cycle, which he flew the next month in Worthington, Ohio, reaching an altitude of 3 000ft.

Nicknamed, 'Bird Boy', by the press, Dixon and his Sky-Cycle became an overnight sensation. He eventually built seven more dirigibles using money he and his mother raised by issuing $10,000 in stock – a considerable sum at the time. But Dixon's invention had a drawback: it was seriously underpowered. It worked well enough on a calm day, but it was impossible to generate enough pedal power to combat anything more than a light breeze. As *Aerial Navigation* noted about one of his flights, 'Mr Dixon found considerable difficulty in returning to his starting point.'

Dirigibles were eventually superseded by aeroplanes, which caused Dixon to switch his allegiance from lighter-than-air to heavier-than-air flight. In August 1911, he received his pilot's licence (only the forty-third to be issued), making him the youngest licensed aviator in the world, or so newspapers claimed. That next month, Dixon achieved another first when he flew across the continental divide, surmounting the previously insurmountable Rocky Mountains.

Sadly, one week later Dixon was dead, killed during a demonstration flight at the Spokane Interstate Fair. His last words shortly before take-off were, 'Here I go!', after which his plane flipped over. He was only 19 years old at the time.

The truth is, Dixon wasn't the first to conceive of the Sky-Cycle,* although his worked best. Nor was he the last. French balloonist Stephane Rousson tried crossing the English Channel in a pedal-powered airship of his own design in 2008. But Rousson ran into the same problem as Dixon – his airship was too underpowered to make the crossing. Proof, if any is needed, that the Sky-Cycle remains White Elephant Technology.

4 The Amphicar

The 1960s were the Age of Aquarius, which is when the Amphicar had its day. A classic hybrid, it was an amphibious car that turned into a boat. It wasn't the first of its kind, or the last, but it's probably the most well known.

The brainchild of German industrial designer Hans Trippel, the Amphicar debuted at the New York Auto Show in 1961. With a top speed of 70mph on land and 7mph on water, it was quickly dubbed the 'Model 770'.

* One of the earliest versions of the Sky-Cycle was patented by Carl Edgar Myers in 1897 and was called the Aerial Velocipede. Myers eventually built a more robust version, which he named the Sky-Cycle. Cromwell Dixon saw Myers demonstrate his Sky-Cycle at the 1904 St Louis Fair and was inspired to build his own version. Myers, like a lot of WETech inventors, was an eccentric. Originally appointed the fair's Director of Aeronautics, he was soon dismissed for 'obnoxious behaviour'. Myers spent most of his time conducting lighter-than-air experiments at his mansion in Frankfort, New York, which he called the 'balloon farm'.

The Amphicar. (Courtesy Lane Motor Museum)

The Amphicar's rear-mounted, British-built, four-cylinder engine produced an underwhelming 43hp. A four-speed manual transmission drove the wheels on land, while a second gear shift (for forward and reverse) controlled a pair of propellers on the water. Because the Amphicar's propellers and wheels could be operated simultaneously, it was able to drive itself out of the water once its wheels touched the bottom. Needless to say, operating its propellers when backing up on land could cause a problem.

All Amphicars were convertibles (to assist in rapid sinking) and sold in only four colours: Beach Sand White, Regatta Red, Lagoon Blue and Fjord Green. Since the US Coast Guard considered it a boat, Amphicars had to be outfitted with navigation lights and a flag. Some even sported a boat registration on the side.

President Lyndon Johnson owned an Amphicar. He was said to enjoy driving unsuspecting visitors around his Texas ranch, concluding the tour by sprinting the car into a lake, claiming his brakes had failed.

Manufactured in West Germany and marketed in the United States from 1961 until 1968, Amphicars sold for between $2,800 and $3,300 depending on the model year. Later models actually sold for less – never a good sign. Sales were predicted to reach 25,000 units, but fewer than 4,000 were sold – the vast majority of them in the United States with Britain second.

Time magazine wrote that the Amphicar 'promised to revolutionise drowning', but that's not really fair. In 1965, eight people in two Amphicars successfully crossed the English Channel (with the top up) before driving 372 miles to the Frankfurt Auto Show.

Production ceased in 1965, but Amphicars continued to be sold in the United States until a change in automotive regulations in 1968 put an end to distribution. The Berlin factory closed that same year. The remaining inventory was eventually bought by Hugh Gordon, who continues to sell replacement parts today.

Why didn't the Amphicar succeed? The car worked more or less as promised, but how many people want a car that drives on water? Not enough to support production at the price at which it was offered. Plus, changes in US regulations eliminated its most important market.

You can still find Amphicars today as approximately 400 operate in seven countries. There's even an International Amphicar Owners Club for enthusiasts (www.amphicar.com).

You can take an Amphicar tour of the waters around Disneyworld in Orlando, Florida.

You can also see an Amphicar at the Lane Motor Museum, Nashville, Tennessee; the Forney Museum of Transportation, Denver, Colorado; the Saratoga Automobile Museum, Saratoga Springs, New York; and the Louwman Museum, Den Haag, the Netherlands.

5 Flying Aircraft Carriers

If underwater aircraft carriers seem counter-intuitive, consider the flying aircraft carrier.

The USS *Macon* (ZRS-5) was the largest, most expensive, most technologically sophisticated aircraft when it was commissioned in 1933. More than two football fields long and fourteen storeys tall, the helium-filled airship was a flying aircraft carrier. Carrying up to five fighter planes in its belly, which it could launch and retrieve in mid-flight, its Curtiss F9C-2 biplanes (named Sparrowhawks) had a skyhook on their upper wing that allowed them to be lowered from the airship on a trapeze. When the pilot was ready to launch, he pulled a lever on the skyhook's stem, releasing his plane into a vertiginous drop. When he wanted to land, he flew beneath the airship, latched his skyhook on the trapeze, and was cranked back on board.

USS *Macon* over Manhattan. (US Naval Historical Center)

Sparrowhawk pilots were so proficient at airship take-offs and landings that none were lost. However, there was at least one instance when someone had to climb down the trapeze outside the airship and hammer on the skyhook's jammed mechanism until it released the aircraft.

Designed in the days before radar, the *Macon* was meant to patrol the Pacific Ocean faster, cheaper and more efficiently than the seagoing cruisers the US Navy relied on. Its mission: to prevent a surprise attack by the Japanese Imperial Navy against Hawaii and California.

The USS Macon inside Hangar One at Moffett Field in Mountain View, California. (US Naval History and Heritage Command)

As if being a flying aircraft carrier wasn't incredible enough, the *Macon* was a self-contained city in the sky with everything the eighty-man crew needed to stay aloft for nearly a week; this at a time when most aeroplanes could fly only a few hundred miles, carried a handful of people, flew largely in daylight and crashed with alarming frequency.

A Curtiss F9C-2 biplane (nicknamed a Sparrowhawk) suspended from the USS *Macon* by its skyhook. (US Naval Historical Center)

The *Macon*'s accommodation included separate sleeping quarters for the officers, chief petty officers and enlisted men; a galley that turned out hot meals three times a day; a mess hall; a sick bay; a weather office; a ward room; a smoking room; and a radio shack. Additionally, it boasted eight machine gun emplacements, eighteen telephones spread throughout the airship and a spy car that could be lowered on a cable to watch the enemy below.

Built by a subsidiary of the Goodyear Tire & Rubber Company in partnership with the same German firm that would build the *Hindenburg*, the *Macon* and its sister ship, the *Akron* (ZRS-4), captured the heart of the American public. Nicknamed the 'Queens of the Skies', the two giants dominated popular culture in the same way the space race would for a later generation.

Unfortunately, neither the *Akron* nor the *Macon* lived up to expectations. The engineer responsible for designing both airships discovered a potentially catastrophic flaw in their airframe but, despite voicing his concerns to the navy, was repeatedly

The catwalk of the USS *Macon*. (US Naval Historical Center)

ignored. When the *Akron* crashed in a storm off the coast of New Jersey in 1933, killing seventy-three of her seventy-six-man crew, it was the worst disaster in aviation history. Two years later, the *Macon* suffered a catastrophic structural failure off the coast of California and sank in the Pacific Ocean.

Dr Karl Arnstein, Goodyear–Zeppelin Chief Engineer. (San Diego Air and Space Museum Archive)

The back-to-back disasters two years before the *Hindenburg* burst into flames put an end to the American rigid airship. Given its WETech pedigree, we're unlikely to see another.

The *Akron* and *Macon*'s only surviving F9C-2 Sparrowhawk can be seen at the Smithsonian National Air and Space Museum's Steven F. Udvar-Hazy Center in Chantilly, Virginia.

Chapter Two

Why Some Things Shouldn't Fly

> Ever tried. Ever failed. No matter. Try again. Fail again. Fail better.
>
> Samuel Beckett

Jeeps, tanks, submarines and, yes, even people all have one thing in common. They weren't designed to fly. And yet WETech inventors have been trying to get them airborne for decades.

An entire book could be dedicated just to flying cars, jet packs, personal helicopters and wing suits. Some of these will be touched on in later chapters, but this one concentrates on stranger inventions that flew with unfortunate consequences.

6 The Flying Tank

In 1932, a prickly American engineer named J. Walter Christie drew up plans for what can only be described as a flying tank – a 5-ton armoured vehicle equipped with detachable wings and a propeller. Christie claimed that, after landing on the battlefield, you only had to pull a lever to release its wings before trundling into battle.

Christie said his tank could move at speeds up to 100mph and only required 100ft for take-off. Nor did it need a conventional runway; it could land just about anywhere – an extraordinary claim for a tank.

'The flying tank is a machine to end war,' Christie told *Modern Mechanics and Inventions*. 'Its possession will be a greater guarantee of peace than all the treaties that human ingenuity can concoct.'

Christie had pitched this previous tank designs to the US Army but, despite his enthusiasm, his flying tank never got traction. That didn't stop the Soviets, the UK and Japan from experimenting.

Both *Popular Science Monthly* and *Modern Mechanics and Inventions* ran illustrated articles about the Christie Flying Tank in their July 1932 editions.

FLYING TANKS ...War's Deadliest Weapon

Side view of the flying tank in course of construction at Linden, N. J. Note the undercarriage is armored

Inside the flying tank. Note 1,000-horsepower motor to drive propeller, wheels, and tread

Trench Warfare Doomed by Tanks That Fly

So bold that it staggers the imagination is the flying war tank which is described in this article. Revolutionary in nature, it may change the whole system of warfare and do away with the trench. Nothing like it exists at present anywhere in the world. It was conceived by a clear-headed engineer who has cut the weight of the ponderous machine until it should be able to get into the air. If the tank is a success, Mr. Christie has said he will present it to the United States Government.

A BATTLE is raging. What seems to be a fleet of attacking airplanes is sighted. Suddenly terror descends out of the sky. Swooping low, the machines are revealed to be armored tanks with wings. They land. The wings drop off, and into action roars a squadron of four-ton tanks, spitting death from three-inch guns.

That is the picture proposed by J. Walter Christie, inventor of what is perhaps the most terrifying war weapon ever conceived—a flying tank. No nightmare of mere fancy is this creation, though its inspiration might have come straight from the pages of one of Jules Verne's romances. The first of these amazing machines was taking shape in a Linden, N. J., factory as this issue of the magazine went to press. To the few who have been permitted to see it, the inventor declared that he would shortly demonstrate his aerial wonder in action.

Even those who might ordinarily consider the idea of a flying tank fantastic hesitate to express doubts of its success in view of the designer's standing. Famous as an automobile racing driver of a quarter-century ago, Christie has become equally noted as an inventor in the field of automotive engineering. It was he who pioneered in perfecting the front-wheel drive for automobiles. He has repeatedly designed successful war tanks for the United States Army, which has purchased and is using a number of them in its motorized branch—notably a remarkable tank with a detachable tread and rubber-tired wheels enabling it to speed at 100 miles an hour on a smooth highway, demonstrated last year. It is around this successful machine that he has designed his flying tank.

Skillful design has pared the weight of the hybrid vehicle to a little more than four tons—less than the weight of many standard airplanes—without sacrificing the armor protection. A single 1,000-horsepower motor drives the air propeller, the wheels, and the tread. Two men constitute the crew. When the pilot lands, he starts the endless tread whirling so that he can come to earth at high speed. Pulling a single lever discards the entire wing structure.

For future development Christie plans a different machine. Instead of a tank with detachable wings, he proposes a detachable airplane, with its own pilot, that could carry a tank to a scene of combat, deposit it while skimming the earth, and return for more tanks without stopping to land.

Military experts foresee that success of the new weapon would render trench warfare obsolete. Hitherto planes have been unable to transport an effective battle force behind an enemy's lines. But with a fleet of winged tanks, a frontal attack no longer would be necessary. There would be no way of stopping a fleet of these formidable engines of war from sailing over the front lines, landing behind the enemy trenches, and turning to charge upon the enemy's rear. Heavy guns capable of stopping the onslaught could not be reversed and pointed rearward in time to be of any real use.

WHEN ARMORED TANKS DROP OUT OF THE AIR

Our artist's conception of the flying tank in action. This amazing war weapon has been fully designed and one is so near completion that demonstration is expected soon

Drawing by R. G. SEIELSTAD

HUNDRED-MILE-AN-HOUR LAND TANK

This fast land tank was designed last year by J. Walter Christie for the U. S. Army and has a speed of 100 miles an hour. Like it, the flying tank will be equipped with detachable tractor treads

12 POPULAR SCIENCE MONTHLY

JULY, 1932 13

The only known photo of the A-40 *Krylya Tanka*, designed in 1940 by engineer Oleg Antonov. The photo is probably of a model.

The Soviets started by dropping lightweight tanks strapped to the underside of a heavy bomber. But the bombers had to fly so low before dropping their tanks they risked being shot down. The next step was turning a tank into a glider. The result was the Antonov A-40, dubbed the 'winged tank'. Named after its designer, Oleg Antonov, it was a T-60 light tank wrapped in a detachable cradle with fabric-covered biplane wings and a tail.

Even though the T-60 was classified as a light tank, it had to be stripped of its armour, and carried minimal fuel and no ammunition during its first drop test. But the tank proved so heavy the bomber towing it aloft had to cut it loose early. Despite the unscheduled deployment, the A-40 successfully glided back to earth, after which its pilot detached the wings and tail and drove the tank back to base. Although the test was judged partially successful, development had to be abandoned because no aircraft in the Soviet arsenal was powerful enough to tow such a heavy tank into the air.

○○○

Raoul Hafner was an Austrian engineer specialising in autogyrcs* when he moved to the UK during the 1930s. Hoping to sell one of his inventions to the military, Hafner was able to

* An autogyro is a type of aircraft that combines overhead rotor blades (used for lift as in a helicopter) with a forward- or rear-mounted propeller (as in a conventional aeroplane). One important difference is that an autogyro's rotor blades are unpowered. In other words, they turn as result of the airflow generated by the craft's forward movement. Although an autogyro can hover like a helicopter, it cannot take off vertically; it needs a runway. Nevertheless, its take-offs and landings require very little distance.

interest the British Air Ministry in his most unusual design: the Rotatank.

The Rotatank was an attempt to marry a 16-ton Vickers Valentine tank with the detachable rotors of a gyrocopter and an aeroplane tail. The Rotatank was to be towed into the sky by an aeroplane until the time of its release, when the tank would 'helicopter' its way to a landing. A dolly with a set of wheels would be used for take-off but dropped once the tank got airborne, enabling it to land on its tracks.

Since no RAF aircraft could tow such a heavy object, the Rotatank never got beyond the design stage. Like Russia, the British developed gliders during the Second World War to carry and/or drop a light tank into battle, rendering the need for a flying tank obsolete. In appreciation for his efforts Britain interred Hafner as an enemy alien, only releasing him when he became a naturalised subject.

Japan also experimented with flying tanks during the Second World War. The Special No. 3 Flying Tank* developed by Mitsubishi in 1943 had detachable wings much like the Soviet version. It could also be transported in a glider towed by a heavy bomber. The Japanese flying tank never went into production, though, because Japan realised it was far easier to push a tank out of an aeroplane with a parachute than fly one into battle of its own accord.

There's no need for a flying tank today given the ability to parachute them into battle. The US has perfected dropping unmanned tanks out of a C-17, but the Soviets developed the neat trick of dropping a tank out of an aeroplane with the driver and gunner inside (no, thank you). A parachute slowed its descent, but the show-stopper was firing a retrorocket as the tank neared the ground. One thing's for sure, you wouldn't want to be underneath one when it lands.

* Called the Maeda Ku-6 Sora-sha (sky vehicle), or the Kuro-sha (black vehicle).

7 The Aerial Rowboat

It's hard to believe that something as unusual as the aerial rowboat was ever invented. It's even more incredible when you realise that two different inventors came up with the same idea in the same year: 1905.

The first aerial rowboat was invented by Thomas Scott Baldwin. Baldwin, who called himself 'Captain', was an important name in the early days of US aviation. Born in 1854, he had a hard-luck upbringing. Orphaned at an early age, Baldwin joined a travelling circus as an acrobat at age 14. Baldwin's circus career involved dangling from a trapeze that hung from a hot-air balloon tethered to the ground. Later, he graduated to parachute jumps from the same balloon. It was this early experience with hot-air balloons that led to Baldwin's interest in finding a way to steer a balloon. The 'dirigibility', or steering, of balloons had never been accomplished in the United States. But after four years of experimentation in Oakland, California, Baldwin built the California Arrow, America's first blimp.

The California Arrow was a 53ft-long, non-rigid dirigible filled with hydrogen. After a series of successful test flights over Oakland in 1904, Baldwin took the California Arrow to the St Louis World's Fair, where he successfully completed what many considered at the time to be the first controlled powered flight in the United States. The Wright Brothers had flown at Kitty Hawk ten months earlier, but since they were highly secretive about their achievement, no one was quite sure what they'd accomplished.

Fresh from his St Louis triumph, Baldwin began demonstrating his now-famous airship at Chutes Park in Los Angeles. The California Arrow flights helped drive record admissions to the amusement park, which is why Chutes gave Baldwin a contract to exhibit a small dirigible of his own design that could be rowed through the air like a boat.

Baldwin's aerial rowboat was a simple craft. Its hydrogen-filled envelope was smaller than the California Arrow. It also had a kayak-shaped frame that the pilot sat in to row the craft. The invention's most significant feature was that it didn't need an engine or a rudder to be steered. Instead, it was propelled through the air by bamboo oars with oversized blades made of silk.

Baldwin, who loved to eat, was too heavy to pilot his aerial rowboat, so he hired a skinny 23-year-old named L. Guy Mecklem to fly his invention. Mecklem called Baldwin's rowboat a 'primitive arrangement', but that didn't stop him from taking it on a trial flight, which proved problematic.

If you look closely at the bottom of the photo, you can see both oars of Baldwin's aerial rowboat with their oversized blades.

'The take-off day arrived and I climbed in the flimsy framework,' Mecklem recalled. 'The sun got hotter and the hydrogen expanded and nothing I could do ... would stop' the rowboat from rising.

Mecklem soon found himself stranded 2,000ft above the crowd with a broken oar and a malfunctioning valve that refused to release hydrogen. The aerial rowboat drifted helplessly until the sun went down, allowing the hydrogen to cool enough for Mecklem to land.

The next day, Baldwin strung a 300ft-long wire between two poles before attaching it to his aerial rowboat with a rope. Mecklem used the guideline, which kept him from drifting away, for rowing practice. Two weeks later, he was ready to perform.

Captain Thomas S. Baldwin, inventor of the aerial rowboat, in 1914. (Library of Congress, photographed by Harris & Ewing)

'On a calm day we could put on a pretty good show,' Mecklem recalled in his unpublished memoir.

Ascending a few hundred feet, Mecklem amused the audience by bombarding them with bags full of peanuts. He also threw his hat overboard, quickly paddling down to retrieve it. His favourite stunt, however, was rowing Baldwin's invention towards a pretty girl in the viewing stand before pulling away at the last second, much to the crowd's amusement.

Baldwin's aerial rowboat could do about 4mph on a calm day but, like Dixon's Sky-Cycle, was powerless in a headwind. When the craft became unmanageable, Mecklem carried a rope with a weight attached that he could drop to the ground like an anchor to assist in his landing.

Baldwin's aerial rowboat came to an abrupt end one night when the hydrogen in its envelope burst into flames, destroying the craft. Though he couldn't prove it, Baldwin claimed someone had thrown a cigarette near his airship with the intention of destroying it. Shortly thereafter, Mecklem quit, claiming Baldwin was too 'overbearing', which isn't hard to believe given the disposition of many WETech inventors.

Baldwin's aerial rowboat lasted only six months, but that was long enough to invite competition. That same year, Alva L. Reynolds launched his own version at California's Fiesta Park, which he called the Man Angel.

Early aviators were half carnival barker, half daredevil act, which is how they drove ticket sales. To prove that 'anybody of ordinary physical ability' could operate his Man Angel, Reynolds allowed 17-year-old Hazel Odell to pilot his craft. When the *Los Angeles Herald Examiner* caught wind of the stunt, it ran an article about it, which helped promote Reynolds's act.

The Man Angel was about 36 ft long and 15 ft in diameter.

The Man Angel.

Reynolds was financially ambitious, so he built six Man Angels, renting them out to advertisers and to fairs and amusement parks around the country. He even operated a flying school to teach people how to row his Man Angel, as well as sponsoring a series of long-distance races against a car – all of which his Man Angel lost.

Eventually, Reynolds fell foul of the law. His Man Angel No. 3 was dropping promotional flyers on a Los Angeles street when a policeman tried to arrest the pilot for littering. The officer charged up five flights of stairs to where the Man Angel was anchored to a rooftop flagpole but the pilot simply rowed away, escaping the not-so-long arm of the law.

After one of Reynolds's Man Angels collided with a tree, his efforts on behalf of the aerial rowboat petered out. Instead, he turned his attention toward other inventions, including generating electricity from ocean waves (which is something people are still exploring today), and a method for keeping barnacles off piers. Meanwhile, Baldwin sold the US Army a 95ft-long blimp for $10,000. Designated the SC-1, it was the army's first powered aircraft.

Reynolds is largely forgotten today, while Baldwin is remembered as the 'father of the American dirigible'. However, for a brief time they both captured the public's imagination by rowing across the skies.

8 The Flying Submarine

As strange as it may sound, an underwater aircraft carrier makes a lot more sense than a flying submarine. But a flying submarine was exactly what Donald V. Reid built and tested in 1964.

Reid was a classic WETech inventor. He built his Reid Flying Submarine or RFS-1 (there was no RFS-2) in his backyard. An inveterate tinkerer, Reid had a history of using whatever he found to make his inventions work. He once borrowed the motor from his wife's Kenmore vacuum cleaner to power a model helicopter. Another time he used the family toaster for the air scoops on a powerboat he built, in which he raced and won a first prize. Even Reid's son, who penned a sympathetic biography of his father, called him eccentric. But Reid showed mechanical aptitude from an early age, taking apart clocks to see how they worked before putting them back together. His day job at the Naval Turbine Test Station in New Jersey meant he was far more skilled at building things than the average hobbyist.

The idea of a flying submarine has been around for a long time, but mostly in the form of science fiction. Flying subs appear in Jules Verne's *Master of the World* (1904) and Percy F. Westerman's *The Flying Submarine* (1912), in addition to television series such as *Voyage to the Bottom of the Sea* (1964) and *UFO* (1970).

The story of how Reid came up with the flying submarine may be apocryphal (many origin stories are), but he claims to have been working on a radio-controlled model sub in his basement one night in 1956 when inspiration struck in the form of model aeroplane wings falling off a shelf and landing on his model sub. Hence, the flying sub was born.

Combining an aeroplane and submarine is a major challenge for several important reasons. A submarine hull has to be rigid and strong enough to withstand crushing pressure. This is in direct contrast to an aeroplane's fuselage, which needs to be light and flexible.

With the help of his family, Reid built a series of progressively larger models to prove his concept. Reid completed his first prototype in 1961. Two years later, he was granted a US patent* for a single-seat floatplane that could land on water, flood its fuselage and sink beneath the surface – something seaplanes only did when they crashed.

The Reid Flying Submarine (RFS-1).

Reid's flying sub was assembled from parts scavenged from junked aeroplanes and incorporated such unusual objects as a steel bedframe and two galvanised rubbish bin lids. When the flying sub was finished, it had a long, cigar-shaped fuselage with wings and a tail. It also sat on a pair of pontoons that could be filled with water when it was time to submerge. A four-cylinder aircraft engine powered the craft in flight, while an electric motor propelled it underwater at a turtle-like 2 knots.

* Patent No. 3,092,060.

The most submarine-looking part of the RFS-1 was a pylon resembling a sub's sail on which the engine and propeller were mounted. Since the pilot sat in the nose of the craft, his head was so close to the spinning prop it looked like it might get chopped off.

Converting the RSF-1 from plane to sub was also complicated. First, the pilot had to remove the propeller before he could submerge. He also had to cover the engine with a rubber seal to keep it dry. Since the RFS-1's cockpit was open to the elements, the pilot had to don a scuba tank and wetsuit whenever the plane submerged, which at its deepest was only 12ft. Sceptics soon called Reid's flying sub 'the flub'.

A full-scale version of the RFS-1 made its first test flight on New Jersey's Shrewsbury River in 1964 with Reid's son, Bruce, at the controls. Sporting dual registration numbers, one for a plane and the other for a boat, the RSF-1 revved its engines, filling the air with a mix of oil and salt spray as Bruce pushed the throttle forward.

What was clear from the test was that Reid's flying sub was seriously underpowered. It barely got airborne, flying no higher than 75ft. Skipping across the waves, the RFS-1 managed only short hops in the air. Submerged, it reached a little more than 6ft in depth.

But despite a shaky start, including an early crash, Reid's flying sub did everything it was supposed to: it flew, submerged and resurfaced in a controlled way. Unfortunately, it did none of these things particularly well.

The Military Invention Board reviewed the RSF-1, deeming it impractical. Unable to find investors, the flying sub was never commercialised. Reid never stopped working on his inventions, though. He died in 1991 aged 79.

If it's easy to make fun of Reid's 'flub', it's important to remember that he achieved something neither the US, British nor Soviet military could despite their repeated efforts: a full-scale, working prototype of an actual flying sub. As a result, he probably deserves more credit than he gets.

Reid's RFS-1 can be viewed at the Mid-Atlantic Air Museum in Reading, Pennsylvania.

9 The Hafner Rotabuggy

The Willys MB, otherwise known as a jeep, was much loved during the Second World War. As it was simple to operate, durable and reliable, the US Army depended upon it to get to places. So, of course, the military had to take a good thing and make it better, with predictably poor results.

Raoul Hafner of Rotatank fame took his love of autogyros and, applying it to a Willys MB, created what is popularly known as the Hafner Rotabuggy. Like his Rotatank, the Rotabuggy was towed into the sky behind an aeroplane. It also had an immense tail and detachable rotor to assist in flying. I guess when you specialise in hammers everything looks like a nail.

Hafner got further with the Rotabuggy than his Rotatank. In 1942, R. Malcolm & Co. Ltd was hired by Great Britain's Airborne Forces Experimental Establishment to build a prototype. A driver used the steering wheel to control the Rotabuggy during take-off. Once it was airborne, a pilot, who sat in the passenger seat next to the driver, took over using a control column to operate the rotors.

The first trial of the 'M.L. 10/42 Flying Jeep', as it was officially known, was conducted in November 1943 when a truck towed it down a runaway. The truck couldn't generate enough speed to get the Rotabuggy airborne, however, so a more powerful vehicle (a Bentley racing car) was brought in. Even when the Rotabuggy was tethered to a racing car it was only capable of short hops. Longer hops were made possible in February 1944 when a Whitley bomber towed the Rotabuggy while taxiing around the airfield.

Tests showed the Rotabuggy was prone to severe vibrations at speeds in excess of 65mph. It was also tail heavy. A jeep is no more aerodynamic than a tank, which is why the Rotabuggy's pilot had difficulty holding the control column as it was towed through the air. There were also a few fraught moments during landing when the pilot returned control to the driver. After one test flight, the pilot is said to have stumbled out of the Rotabuggy and lay down on the runway, where he remained for some time. The account can't be verified, but it has the ring of truth about it.

The Hafner Rotabuggy, showing the side view. (Courtesy Ian Tong/Flickr)

Hafner made improvements but a heavy landing during one of the test flights damaged the 'aircraft's' rotor blades, grounding it until they could be replaced. Finally, in September 1944 a bomber towing a Rotabuggy took to the air. After being released, the Rotabuggy flew for ten minutes at an altitude of 400ft and a speed of 65mph. The test was judged unsatisfactory, in part because the bomber had to fly so slowly while towing the Rotabuggy it almost stalled.

Like most inventors, Hafner was determined to solve the problems. His efforts were nipped in the bud, however, when it was decided gliders were a cheaper, easier means of carrying a jeep into battle. And so the Rotabuggy was cancelled before reaching production.

An original Hafner Rotabuggy is rumoured to be in storage somewhere in England awaiting funds to be restored. In the meantime, a replica can be viewed at the Museum of Army Flying at Middle Wallop, England.

10 The Perkins Man-Carrying Kite

Samuel F. Perkins was a Harvard graduate and Boston-based kite maker who conducted experiments with 'man-carrying kites' between 1910 and 1930.

The principle of the Perkins Man-Carrying Kite was simple. A lead kite, 18ft high, was flown into the sky to test wind conditions. If the wind proved strong enough, a series of six to seven stringer kites followed on the same line until there was enough lift to raise a man sitting in a bosun's chair high into the sky. Meanwhile, a dedicated ground crew operating a winch reeled the man in or out depending on the weather.

Perkins began his career assisting Thomas C. Baldwin (just like Guy Mecklem) in his country fair demonstrations. When strong winds prevented Baldwin's dirigible from performing, Perkins used his kites to entertain the crowd.

Perkins got his big break at the 1910 Harvard-Boston Aero Meet. The first aviation meet held on the east coast, it was such a big deal the *Boston Globe* offered $10,000 for the fastest flight from Atlantic, Massachusetts, to Boston Light and back. Sixty-seven thousand people turned out, expecting to see 'aeroplanes' streak across the sky. Instead, newspapers reported the biggest spectacle was watching Perkins be lifted 200ft off the ground by five huge man-carrying kites.

Perkins's fame spread rapidly after that. It didn't hurt that he flew reporters in his invention, proving he wasn't just an inventor but a savvy publicist, too. Perkins may have been ambitious, but he wasn't crazy. In addition to Harvard, he attended the Massachusetts Institute of Technology (MIT), which gave him a scientific grounding for his aerial experiments. The kites themselves were of simple construction. Their frame was made of spruce wood covered in silk. Once enough lift was generated, the aeronaut would sit in a swing-like contraption and be winched into the sky for up to a mile. Subsequent versions of the Perkins kite were said to lift up to 2,000lb with flights lasting ninety minutes.

The year 1910 proved a watershed one for the Harvard graduate. That's when his kites lifted him to an altitude of 300ft, setting what was believed to be a world record. By 1911, Perkins was regularly landing on the front page of US newspapers. The *Los Angeles Herald* even called him 'the greatest authority in the world on man-carrying kites' – undoubtedly, a small group.

Perkins racked up a number of firsts, including a 385ft altitude record and the first wireless message sent from a kite. His kites were also used for aerial observation, photography and as a platform for sharp shooting. He even lent his kites for target practice, though the pilot remained on the ground.

Kite demonstration at Dexter Field, Providence, Rhode Island. (Courtesy Boston Public Library, Leslie Jones Collection)

Like most inventors, Perkins was eager to commercialise his invention, which is why he promised reporters they would see him 'do stunts never seen before'. In the meantime, he supported himself by selling billboard space on his kites. He also got a cut of the gate receipts at the fairs, amusement parks and aviation meets where he performed. His biggest wish, though, was to sell his kites to the military.

Perkins's first opportunity came in January 1911 when his kites lifted a US Army Signal Corps officer over Los Angeles. Two weeks later, Perkins flew members of the Thirtieth Infantry at the Tanforan Aviation Meet near San Francisco. When Rear Admiral Pond saw Perkins's kites perform, he asked whether an observer could be sent over water. Quick to seize an opportunity, Perkins replied yes.

Samuel F. Perkins demonstrating his kite on Broad Street, Philadelphia, 10 November 1910. (Library of Congress)

The next month, Lt John Rodgers, a famous aviator in his own right, was strapped into one of Perkins's kites and sent 400ft above the deck of the USS *Pennsylvania* (ACR-4) as it steamed towards San Diego. But the US Navy was not a buyer, at least not in volume. As Perkins later explained, 'No money was available ... (so) the Navy was left for future consideration.'

The lack of a sale was a disappointment, but Perkins's accomplishments were so highly regarded he was invited to England to demonstrate his man-carrying kite at the festivities for King George V's coronation.

However, not everything went smoothly for Perkins. In November 1910, he had what one newspaper called a 'slight mishap', when the rope connecting his kites snapped, causing him to plummet 75ft to the ground. Perkins used his stringer kites to break his fall, avoiding serious injury, but it was a close call. It would be the first of many such near misses, even after he replaced the rope his kites relied upon with steel cable.

A few weeks later, Perkins experienced yet another mishap, this time over Kansas City when a cyclone unexpectedly swept in and caused him to plummet 150ft to the ground. Perkins used his stringer kites to check his fall, but it was a close call. Next, he was 200ft in the sky at a Los Angeles event when a biplane severed the cable anchoring him to the ground. Although three of his kites were destroyed, Perkins used the remaining ones to parachute to safety, a trick that saved his life.

The First World War brought renewed interest in Perkins's kites. Both Germany and France used kites for scouting, so the US Signal Corps ordered another round of tests, but once again nothing came of it.

One problem with the Perkins Man-Carrying Kite was stability. Getting his invention into the air was easy, but if the wind

A camera man lifted by man-kite at Brockton Fair. (Courtesy Boston Public Library, Leslie Jones Collection)

shifted the kites could veer out of control. Another problem was that kites make for an easy target, something that didn't endear them to an occupant suspended over a battlefield. Additionally, improvements in aeroplanes and observation balloons soon rendered the man-carrying kite obsolete.

Perkins kept filing patents, refining his kites and conducting demonstrations, but he was unable to persuade the military to buy his product. He did have some success, though. Lt John Rodgers (who'd flight tested a Perkins kite for the navy) took one of his signal kites with him on his 1925 attempt to fly non-stop from San Francisco to Hawaii. When the army sent five amphibian planes on a goodwill tour of South America, they too carried a Perkins signal kite. Even famed polar explorer Admiral Byrd took three Perkins signal kites with him when he departed on his first Antarctic Expedition in 1928.

Sadly, Perkins's invention largely disappears from the historical record after 1930. Following a long illness, the 72-year-old pioneer died near Dorchester, Massachusetts, on 7 October 1956.

Though the Perkins Man-Carrying Kite is a prime example of White Elephant Technology, its legacy can still be found today. Just look into the sky at an ocean resort and you'll probably see a tourist flying in a man-carrying kite. Although this is not the use Perkins envisioned, he'd probably be happy his invention survived him.

Chapter 3

Inventive Deaths

> Success is not final. Failure is not fatal. It is the courage to continue that counts.
>
> Often (but incorrectly) attributed to Winston Churchill

Inventions are unproven until tested, and usually the first person to test them is the inventor. That's when the trouble starts. Inventing has enough dangers, including financial loss and crushed egos, but when a zealot's belief clouds an inventor's judgement bad things can happen. What follows is a brief survey of five overly optimistic inventors whose enthusiasm led to their demise.

11 Franz Reichelt and His Wearable Parachute

One of the defining characteristics of a WETech inventor is the intensity of their conviction. Franz Reichelt proved this when he jumped off the Eiffel Tower wearing a parachute of his own design.

Reichelt was a successful dressmaker who'd emigrated from Bohemia to France. Concerned about the high death toll of pilots during the early days of aviation, he may also have been motivated by the 10,000 francs prize offered by the Aéro-Club de France for the first person to design a workable parachute weighing less than 55lb.

Many people are credited with inventing the parachute, including Leonardo da Vinci. But by the time Reichelt came along the French were the leaders in parachute design. The first man credited with a successful parachute jump was Louis-Sebastien Lenormand in 1783.* Lenormand's invention was intended to help people escape a burning building. Given its kite-like rigid frame, it's not what we'd recognise as a parachute today. The honour falls to André-Jacques Garnerin, who jumped from a hydrogen-filled balloon over Paris using a silk parachute in 1797.

Wanting to save lives, Reichelt set about sewing a parachute of his own design to be worn like a suit. During a fall, the wearer extended their webbed arms, increasing the suit's surface area, and this, along with an overhead canopy, would slow their descent.

Reichelt tested several versions that looked similar to today's wing suits. His early experiments involved dropping a mannequin wearing his device from the fifth floor of his apartment building. Reichelt even wore one himself, jumping into a pile of hay from a height of 30ft.

But the more Reichelt modified his design, the more problems he ran into; mainly, his parachute worked *less* well. At

* Others may have parachuted before Lenormand, but the historical record is spotty.

Franz Reichelt.

one point, he even broke his leg after making a jump wearing his invention.

Reichelt felt the problem could be solved if his wing suit was dropped from a height high enough that wind resistance slowed its fall. Finally, after a year of lobbying authorities, Reichelt received permission to drop one of his parachute-clad mannequins from the Eiffel Tower.

On a cold, windy morning in February 1912, Reichelt stood at the base of the famous tower, modelling his wearable parachute for newspaper photographers and newsreel cameramen. Wearing a cap and a thick moustache that curved upwards like a pair of wings, the 34-year-old tailor told the press he was confident his apparatus would work.

However, when he began climbing the stairs to the Eiffel Tower's first platform 187ft above the city, it was clear Reichelt wasn't going to use a mannequin to test his invention but jump wearing it himself. Those present tried talking him out of it, but to no avail. Standing on a chair placed upon a desk next to the platform's railing, Reichelt needed time to work up his courage. With one foot balanced on the edge, he spent forty seconds leaning forward then back, forward then back, before jumping into the abyss. His last reported words were '*Á bientôt*', which translates as, 'See you soon.'

Reichelt leapt feet first, his parachute trailing above him like Superman's cape (and we know how Edna Mode feels about capes). Unfortunately, the only thing he achieved that day was lending an onomatopoeic meaning to the Eiffel Tower. Reichelt's mark upon history was an impact hole 6in deep. Still, you have to admire his determination.

12 Henry Smolinski and His Flying Car

The US auto builders in Detroit made a lot of styling mistakes during the 1970s, but none quite as egregious as the Ford Pinto. But that didn't stop it from becoming the key component in a flying car.

Developed by California's Advanced Vehicle Engineers (AVE) and named for a star in the handle of the Big Dipper, the AVE Mizar married a Ford Pinto to a Cessna Skymaster. If looks could kill, this flying car was born to murder (see cover photo for details).

Its inventor, Henry Smolinski, was a structural engineer with a background in jet engines, aircraft design and missile development. His idea was to take a conventional automobile and integrate it with a small aeroplane so that a person could drive to an airport, fit the car with a set of wings and an engine and take off from the runway. After landing, the pilot/driver detached the car from the wings and engine and drove away.

Smolinski told reporters, 'Our plan is to make the operation so simple that a woman can easily put the two systems together – or separate them – without help.'

In a promotional film, a sceptical-looking woman watches as a man in a very 1970s cappuccino-coloured jacket backs a white, two-door Pinto into a modular airframe consisting of wings and a pusher engine. Four high-strength locking pins were used to make sure the four-seat Pinto remained attached. Nothing could be simpler.

Smolinski's Pinto was modified so the driver controlled the wing's ailerons by turning the steering wheel right or left. Pushing the wheel forward or pulling it back controlled the elevators on the tail, while floor pedals near the accelerator and brake controlled the rudder.

The Pinto's petrol tank would soon have a reputation for exploding when rear-ended, but that was the least of Smolinski's problems. An ill omen came during an early test flight when the attachment holding the Mizar's right wing strut failed, forcing the pilot to make an emergency landing. After coming down in a bean field, the pilot drove the Mizar back to the airport without any trouble.

But the Mizar's most memorable act of violence was to disintegrate in mid-air during a 1973 test flight. Shortly after take-off, the Mizar's right wing folded, sending the car plummeting to earth. As pieces rained from the sky, the flying car hit a tree top before landing on a parked pick-up truck and bursting into flames. Smolinski and his business partner, Hal Blake, were killed instantly.

Obviously, there were problems with the AVE Mizar. One was that the car was overweight, which stressed the airframe. Another was that a bad weld caused the right wing to fail. The AVE Mizar was to be sold at the affordable price of $15,000. Sadly, it cost Smolinski and Blake a whole lot more.

○○○

There's a long history of flying cars, some of which worked better than others. Glenn Curtiss is said to have designed the first one, the Curtiss Autoplane, in 1917. The Autoplane had three wings and a huge, rear-mounted propeller. Curtiss's flying car was a hit when it debuted at the New York Pan-American Aeronautical Exposition. It even managed a short flight, though some describe it as more of a hop than something sustained.

In 1926, Henry Ford also got into the flying car business when he unveiled a single-seat monoplane he called the 'sky flivver'. Ford intended his experimental aircraft to become the Model T of the sky.

'Mark my word,' he told the press, 'a combination aeroplane and motor car is coming.'

But when a sky flivver crashed two years later, killing its pilot, testing ceased. Ford continued to evangelise flying cars throughout the 1940s but never built another one, which is a shame given his genius for low-cost, high-quality production.

13 Alberto Santos-Dumont and the Dream of Flight

Few North Americans know who Alberto Santos-Dumont is, but he was every bit as famous as Charles Lindbergh in his day, maybe more so since most people thought him the first person to achieve powered, navigable flight.

The scion of a wealthy Franco-Brazilian family, Santos-Dumont was born to fly. Only 5ft tall and weighing less than 110lb, he was light as a feather. Fascinated by all things mechanical, he took special pleasure as a boy in the machines on his family's coffee plantation. When aged 7, he was driving a steam-powered tractor. By 12 he was riding in the cab of a Baldwin locomotive as it carried his family's coffee beans to market.

Inspired by Jules Verne's *Five Weeks in a Balloon*, Santos-Dumont grew up admiring the Montgolfier brothers. He even built his own miniature hot-air balloon. But it wasn't until he was 18, when he moved to Paris to pursue his studies, that he took his first balloon ride. Sailing above the rooftops, he enjoyed a high-altitude lunch of cold roast beef, hard-boiled eggs, cakes and champagne. By the end of the flight he was hooked.

Santos-Dumont built his own balloon shortly thereafter, naming it *Brazil* for his home country. Its basket was so small there was barely room for him. When deflated, Alberto carried the balloon around Paris in a valise.

Santos-Dumont was one of the most colourful Lighter-Than-Air (LTA) enthusiasts of his time, which is saying something given the field was filled with eccentrics. An inventive, if restless soul, he was equal parts flamboyant, sensitive and shy. Possibly gay, more likely asexual, he was never known to have a romantic liaison. Nevertheless, no one could match him for charm, grit and flair.

Backed by his family fortune (it helps for a WETech inventor to be rich), he pursued the holy grail of navigable flight by adding an engine to his hot-air balloon.* After learning from these experiments Santos-Dumont graduated to dirigibles in 1898, making his first flight in a non-rigid airship he designed himself.

Santos-Dumont's No. 1 dirigible had a sausage-shaped balloon and a propeller driven by an internal combustion engine he'd salvaged from a three-wheeled vehicle. For a gondola, he relied on ballooning's traditional wicker basket suspended by a few taut cables from the overhead envelope.

A huge crowd gathered in the Jardin d'Acclimatation to watch Santos-Dumont's hydrogen-filled dirigible make its first flight. But as the No. 1 slowly began to ascend it failed to clear the park's trees, bringing an abrupt end to its maiden voyage.

More mishaps followed. In 1899, his No. 2 dirigible began sagging in the middle before a gust of wind blew it into the trees. When the envelope of his No. 5 suddenly burst, stranding him on a hotel rooftop, his determination to succeed remained undiminished.

---

* The French noun *dirigeable* is derived from the French for 'to steer'.

Portrait of Alberto Santos-Dumont.

Santos-Dumont failed several more times, but his airship design was improving. By now he was using an open air catwalk suspended beneath his airship's envelope, which allowed him to slide a weighted bag towards the bow to tip its nose downwards when he wanted to descend, or its stern when he wanted to rise.

Twenty-two days after his No. 5 debacle, Santos-Dumont launched his now famous No. 6. A prominent oilman, Henri Deutsch, was offering a 100,000 francs prize for the first aeronaut to fly a 7-mile round trip course from the Aéro-Club de France to the Eiffel Tower and back in under thirty minutes. Santos-Dumont was determined to rise to the challenge.

His first two attempts ended in failure, but on 19 October 1901 Airship No. 6 took off for the Eiffel Tower. Surrounded by grey skies, *le petit Santos* ascended to an altitude of 1,000ft where, chugging along at 15mph, he headed for the Eiffel Tower as spectators cheered him on.

The pint-sized aeronaut worried that cross winds might blow him against the tower, but his airship circled the monument without mishap. On his return trip, he crossed the Seine, flew over Longchamps racecourse and reached his point of departure twenty-nine minutes and thirty-one seconds after he'd left. When he asked the crowd, 'Have I won?' they roared back, 'Yes!' Santos-Dumont gave half the prize money to Paris's poor to acknowledge his socialist leanings.

One thing that makes Santos-Dumont's 7-mile, thirty-minute flight so impressive is that he undertook it two years *before* the Wright brothers flew at Kitty Hawk. By comparison, Wilbur and Orville's straight-line dash went only 852ft, lasted less than a minute, never exceeded an altitude of 14ft and ended in a crash.

Santos-Dumont's Airship No. 1.

Le petit Santos built eight more dirigibles after that, each one better than the last. Despite forced landings and hair's-breadth escapes including ditching in the Mediterranean, he continued making progress.

Meanwhile, Parisian society fell in love with *le petit Santos*. They especially enjoyed watching him fly over their rooftops in

Santos-Dumont circling the Eiffel Tower in his No. 5 dirigible in July 1901. He failed to complete the round trip within the allotted time, but succeeded two months later.

his No. 9 'runabout'. Used to run daily errands, the No. 9 dirigible was about a third the size of the prize-winning No. 6, making it easier to handle. As Santos-Dumont recalled in his autobiography, *My Air-Ships: The Story of My Life*, it wasn't unusual for him to fly to the Cascade Café, land in a nearby park and, after tying up his dirigible, join his friends for lunch. When he was finished, he'd fly back to his residence on the Champs-Élysées, where two servants grabbed the No. 9's guide rope while Santos-Dumont disembarked, 'to my apartment for a cup of coffee'.

Santos-Dumont dressed impeccably for each flight. Wearing a suit and tie in the fashion of the day, he favoured high, white, heavily starched collars and a soft-brimmed hat that complemented his soulful countenance. Santos, as he preferred being called, was not a handsome man. In fact, he was rather homely. His droopy moustache artfully disguised an over bite, and he was so thin as to appear frail. But his combination of pluck, fortitude and *savoir faire* endeared him to Parisians. One night, while celebrating a flight at Maxim's, Santos-Dumont complained to his friend Louis Cartier that it was too dangerous to take his hands off his airship's controls to check his pocket watch. Not long after, Cartier created the world's first wristwatch specifically to solve Santos-Dumont's problem.

Santos-Dumont's fascination with all things mechanical didn't stop at dirigibles. He also experimented with automobiles and powerboats. But his restless nature led him to shift interest from airships to aeroplanes. Cannibalising his No. 14 dirigible, he built the *14-bis* in 1906, making the first controlled, powered, heavier-than-air (HTA) flight in Europe. The following year, Santos-Dumont set a world HTA record by covering 722ft in less than twenty-two seconds while flying his plane facing backwards.

Santos-Dumont retired from aviation in 1910. Prone to depression, he was rumoured to have suffered a nervous breakdown and secluded himself at home. When the First World War broke out, he blamed himself for promoting an invention now used to kill people. Just the thought left him 'in a flood of tears'.

Santos-Dumont became so depressed he eventually stopped eating. Losing weight he could hardly spare, he checked into one sanatorium after another, seeking a cure. By 1922, his eyes were so hollow, his cheekbones so pronounced, he looked like a skeleton. Finally, Santos-Dumont's nephew persuaded him to visit Brazil, where he was considered a hero, hoping it would reverse his decline.

Unfortunately, the opposite happened. As Santos-Dumont's steamship entered Rio's harbour, a dozen of Brazil's scientific, political and intellectual elite boarded a seaplane (named the *Santos-Dumont*) to greet him. But as *le petit Santos* watched the aircraft come in for a landing, it plunged into the sea, killing everyone on board. Convinced he'd caused their death, Santos-Dumont sunk deeper into depression.

Santos-Dumont and the *14-bis*. (Bibliothèque Nationale de France)

Santos-Dumont spent his final years in Brazil suffering from multiple sclerosis. A 1932 portrait shows a haunted man. Sadly, he lived long enough to witness the Brazilian air force bomb its own people. Driven by despair, he spoke his last words, 'What have I done?' to a lift operator before hanging himself in his hotel room three days after his 59th birthday.

Most Brazilians consider *le petit Santos* the father of aviation. Not only did his face once appear on Brazil's 10,000 cruzeiros note, but more streets, avenues and plazas bear his name than can reasonably be counted. It's a pity more people don't know his story.

14 'Mad' Mike Hughes and His Flat Earth Rocket

On a more contemporary note, Mike Hughes was killed when his home-made, steam-powered rocket plunged back to earth after a failed launch in 2020.

Hughes was a self-taught engineer with very little money who built rockets in his garage with the help of a friend. A shrewd self-promoter (an important skill for WETech inventors), Hughes used his colourful personality to raise money to fund his stunts. After covering his body in bubble wrap, he successfully jumped a stretch limo 103ft in 2002, earning him a Guinness World Record.

Hughes claimed he was more daredevil than inventor. It can be difficult to distinguish between the two, but he had a point. There was nothing revolutionary about his steam-powered rocket other than the fact he built it himself.

In his garage at home.

Where he lived alone.

With his four cats.

Hughes may have been an amateur, but his 2020 rocket launch wasn't his first rodeo. In 2014, he successfully launched the X-2 SkyLimo in Arizona with himself as payload. After reaching an altitude of 1,300ft, the rocket landed safely if not softly. Afterwards, Hughes told the Associated Press he was relieved he hadn't 'chickened out'.

Four years later, Hughes launched his second rocket. This one flew over California's Mojave Desert, reaching a peak height of 1,875ft. After a harder than anticipated landing, Hughes had to use a walker for a month.

The media made fun of Hughes's belief the earth was flat. Indeed, he stencilled 'Flat Earth Research' on one of his rockets. But the truth was a little more complicated. 'People ask me why I do stuff like this,' he explained to the *Los Angeles Times*. 'Basically, it's just to convince people they can do things extraordinary with their lives ... maybe (it) will inspire someone.'

Perhaps, but Hughes's rocket launches were primarily designed to raise money for his ultimate daredevil trick: a part balloon, part rocket he called the 'Rockoon'. As Hughes imagined it, a balloon would lift him in a rocket 25 miles into the atmosphere before separating. The rocket would then take him 62 miles above the earth where space begins, after which it would parachute back to earth.

'Mad' Mike Hughes and his Flat Earth Rocket. (ZUMA Press, Inc./ Alamy Stock Photo)

Before Hughes could build the Rockoon, he had to launch his third rocket intended to reach a mile-high altitude. To be clear, this height, 5,500ft, barely qualifies as 'flight' given commercial airliners regularly fly at 30,000ft. Heck, 'Lawn Chair Larry' reached 16,000ft sitting in a flimsy lawn chair lifted by helium balloons.* Still, 5,500ft wasn't bad for a home-made rocket.

What's clear is that 'Mad Mike', as the media dubbed him, knew what he was doing. 'It's a dangerous thing,' he told Space.com. 'Anything could be catastrophic.' Hughes's rocket was basically sound. An internal tank made of Nitronic 40 stainless steel was

* Lifted by forty-five colourful balloons attached to a patio chair, Larry Walters made a forty-five-minute flight over Los Angeles in 1982. He was spotted by two commercial airliners, after which he used a pellet gun to pop his balloons, allowing him to return more or less safely to earth.

filled with 95 gallons of tap water. When warmed by an immersion heater to 400°F it created 425psi of pressure.

Unlike a liquid or solid fuel rocket, which builds thrust as it climbs, a steam-powered rocket releases all of its thrust upon lift-off. In other words, it's like being shot out of a cannon – the initial momentum is what drives you to apogee.

Hughes's rocket was simple if improvised. Its water tank leaked, causing the pressure to drop. After several scrubbed launches, the 64-year-old daredevil was finally ready. On a sunny Saturday in Barstow, California, he climbed a ladder to reach his rocket's cockpit and scooched inside. The red, white and blue missile, tilted at a 79° angle, looked like something Evel Knievel might have ridden. What happened next was captured on film by the Science Channel.

When Hughes hit the ignition button, a cloud of steam shot out of the craft's rear nozzle, making a sound like a bottle rocket taking off. But as 425lb of thrust drove Hughes's rocket into the sky, its parachute deployed prematurely, ripping to shreds.

Hughes fell short of his 5,500ft goal, but he did achieve terminal velocity on his return trip. Tracing a gigantic arc, he dived back to earth with no means of slowing his descent. The entire trip from blast-off to landing took less than a minute.

The report by the San Bernardino County Sheriff's Department is a fitting eulogy. Called to the launch site at 1:52 p.m., they pronounced Hughes deceased shortly thereafter. According to one eyewitness, there wasn't enough of Hughes left to identify – a sad end to a strange story.

15 Thomas Midgley and His Ambulatory Device

There's no doubt Thomas Midgley Jr was a highly distinguished inventor, even though three of his inventions turned out to be lethal.

Midgley was a chemist with 117 patents to his name. While working for an Ohio-based company in 1921, he discovered a method for eliminating engine knocks by adding tetraethyl lead to petrol. The resulting product, Ethyl Gasoline, made him a fortune, but Midgley's experiments with the additive made him so sick he was forced to take time off from work to recover – a harbinger of problems to come.

Midgley's next big challenge came in 1930 when he began looking for an odour-free, non-toxic, non-flammable gas that could be used as a coolant in refrigerators and air conditioners. He eventually settled on a fluorocarbon he branded and sold as Freon.

During a famous presentation Midgley made before the American Chemical Society, he inhaled a lungful of Freon to demonstrate its safety. When he exhaled it on a lit candle, extinguishing the flame, he proved to his audience in the most dramatic way possible that Freon was neither poisonous nor flammable.

Thomas Midgley.

Midgely received many honours during his lifetime, including election to the National Academy of Sciences, four medals from the American Chemical Society and being inducted into the National Inventors Hall of Fame. However, the lead expelled by Ethyl Gasoline was found to be poisonous, and Freon to deplete the earth's ozone layer, allowing harmful UV rays to penetrate the atmosphere. Subsequently, both were banned from commercial use.

Sometimes, inventors invent something because they want to help people; other times it's because the problem personally affects them. Such was the case with Midgley after he contracted polio at age 51.

His legs now paralysed, Midgley turned his mind to creating a complex system of ropes and pulleys to make it easier to lift him out of bed. But Midgley's invention, like Ethyl Gasoline and Freon, proved lethal. One day in November 1944, he accidentally became entangled in his device and was strangled to death. Allegations that Midgley used his device to commit suicide have never been proven.

ooo

As Zig Ziglar, the motivational speaker, once said, 'It's not how far you fall, it's how high you bounce back.'* Not bad advice for most WETech inventors. Just don't tell Franz Reichelt, who jumped off the Eiffel Tower wearing a wing suit.

* With permission from Zig Ziglar, Zig Ziglar, Inc., www.ziglar.com.

Chapter Four

Freezing Your Ass Off: Exploring Harsh Environments

> Crazy is often being right, too early.
>
> Common saying

Complexity does not do well in harsh environments. Time and again the simplest technology proves best So why do WETech inventors repeatedly introduce complex mechanical systems into extreme environments with predictable results? You can decide for yourself following this brief sampling of overland vehicles that explored the land of failure.

16 & 17 **LeTourneau's Sno-Freighter and Sno-Train**

During the Cold War, the US Air Force needed to build seventy-eight radar stations to warn against a Soviet nuclear strike. What's more, they needed to build them in a hurry. To be most effective the radar stations had to be located in the most remote part of northern Alaska. In fact, the area was so removed from civilisation there were no roads or rail connections and few air-strips for delivering supplies.

The surrounding terrain was rough enough that a single supply trip could take three months. Meanwhile, temperatures dropped to -68°F in the winter. The question, then, was how to transport the men and materials necessary to build the DEW (or Distant Early Warning) radar stations.

The solution was a new kind of off-road vehicle purpose-built to navigate harsh terrain. Designed by Robert G. LeTourneau and his Texas-based company, LeTourneau Technologies, the Sno-Freighter turned out to be one of the strangest-looking vehicles anyone had ever seen.

The Model VC-22 (its official designation) was almost as long as a football field with a locomotive cab and five powered trailer cars. Operated by a four-man crew, the twenty-four-wheeled vehicle was driven by twin 400hp Cummins diesel engines, and could carry up to 150 tons of cargo. If crossing Alaska's frozen tundra wasn't hard enough, LeTourneau's Sno-Freighter could also ford rivers up to 4ft deep.

LeTourneau was an interesting character. A cross between the larger-than-life oil well firefighter Red Adair and a Christian evangelist, he had extensive experience building heavy equipment for the private sector. He also created such exotic military vehicles as a mobile launcher for the Corporal missile (the first nuclear-tipped, surface-to-surface guided missile) and a huge mobile crane designed to remove incapacitated bombers from a runway in a hurry.

LeTourneau's Model VC-22 Sno-Freighter. (R.G. LeTourneau The Man, Machines, and Mission Collection)

LeTourneau built the Sno-Freighter in 1955. After being shipped in pieces to Circle City, Alaska, it was assembled and in the field in less than six days. One of the Sno-Freighter's most notable features was that each of its wheels could be driven independently of the others – a revolutionary innovation that helped make LeTourneau wealthy. The Sno-Freighter's tremendous size also gave it an enormous tyre area to vehicle weight ratio, providing superior traction when traversing the Alaskan wilderness.

The Sno-Freighter performed well in its first season. During its second winter, however, it was involved in an accident, partially burned and had to be abandoned. Subsequently, Alaska Freight Lines, the company that leased the Sno-Freighter to deliver DEW construction supplies, went bankrupt.

But that wasn't the end of LeTourneau's 'land trains', or of the DEW's pioneering supply line. In fact, it was just the beginning. The Army Transportation Corps was impressed enough with LeTourneau's Sno-Freighter that it ordered an upgrade, the Logistics Cargo Carrier, or LCC-1, fondly referred to as the Sno-Train.

LeTourneau (right) travelling out in the field – the Greenland Ice Cap – to check the machine's operation. (R.G. LeTourneau: The Man, Machines, and Mission Collection)

The Sno-Train's control cab was split into two articulated compartments: the front compartment had the drive controls and bunks for the crew while the rear section contained a 600hp Cummins engine, generators and fuel tanks. The diesel engine delivered electricity to each one of its sixteen individually powered wheels, and since the Sno-train was designed to run twenty-four hours a day, the three-man crew slept in shifts.

Although the LCC-1 could not transport as much cargo as the Sno-Freighter, it was an improvement in that it was lighter with a lower centre of gravity. The 10ft-high wheels also distributed the weight more evenly, which helped prevent the vehicle getting bogged down.

LeTourneau delivered his Sno-Train to the army in 1956. After thorough testing it was deployed to Greenland, where it supplied the DEW radar stations from 1956 to 1962. Often knee-deep in snow, it was truly a white elephant.

The Sno-Train (LCC-1) wound up in Alaska, where it was eventually abandoned and ended up in a salvage yard for many years. Fortunately, it was rescued and is now on display at the Yukon Transportation Museum in Whitehorse, Canada.

18 The Rhino

Elie P. Aghnides was a Greek–American inventor who made a fortune placing a tiny metal screen inside water taps and shower heads. Not only did the screen remove chlorine, making water more palatable, but also the added air made soap lather faster. Aghnides's 1943 improvement resulted in his invention becoming standard in nearly every household in the developed world.

Born in Istanbul and raised in England, Aghnides received an engineering degree from the University of Belgium before becoming a naturalised American citizen. He held many patents over the course of his life, but none as odd as the one for the Rhino. An all-terrain vehicle designed during the 1950s to patrol the inaccessible parts of Alaska and Canada, the Rhino combined the speed of a car with the power and stability of a bulldozer to defend against communist attack.

The vehicle, painted bright yellow, was 19ft long, 9ft wide and nearly 10ft high. The driver sat in a bubble above the cab while a passenger (presumably a gunner) sat behind him, facing backwards. The Rhino's secret sauce was a pair of massive front wheels shaped like a globe with ribbing for better grip. Each one of the front wheels was 6ft in diameter and weighed 1,500lb despite being hollow – two thirds of its 5-ton weight. The huge surface area of the two front wheels provided better traction, allowing the Rhino to wade through deep mud. The front wheels were also slanted inwards and this, combined with a low centre of gravity, made the Rhino extremely stable. It could even climb a 65°F angle without tipping over, or so Aghnides claimed.

The Rhino had a similar though much smaller set of wheels in the rear, which made it look like a weightlifter – all bulging chest with a tiny waist. Still, it was extremely versatile. It could reach speeds on land up to 45mph and, since its front wheels

A close up of the Rhino, showing its front wheels.

were hollow, float on the water. Using a Kermath Marine Hydro Jet that swivelled for steering, it propelled itself through water at a snail's pace of 5mph. This means the Amphicar could beat it in a race, whether on land or water.

The Rhino's shell was made of aluminium welded to a steel frame; its 110hp engine similar to cars of its day. Aghnides applied for a patent in 1953. The next year, he hired the Indianapolis-based Marmon-Herrington Co. Inc. to build two prototypes as proof of concept demonstrators.

The Rhino had been conceived as a replacement for tanks, so Aghnides hoped to sell it to the military. Designed to penetrate dense forests and inhospitable swamps, the Rhino could travel through sand, mud, snow and water. It could even mow down small trees in its path. You can watch footage of it in action on YouTube, but it's interesting to note the promotional film cuts away every time a problem seems to arise.

Aghnides intended a future version of the Rhino to have more horsepower, bigger cargo space and armour plating. However,

The Rhino.

the military passed on its purchase, claiming the Rhino's front wheels were vulnerable to attack.

Aghnides was so proud of his invention he kept an ashtray in the shape of its wheels in his luxurious Manhattan apartment. He built a successor to the Rhino with similar-shaped wheels called the Cyclops. But when he fell out with Marmon-Herrington over its construction he ended up suing the company. The court found in Aghnides's favour, ruling 'the vehicle, as constructed, was a failure and totally unfit for demonstration to prospective manufacturers'. For his trouble, he was awarded $120,505.40.

Aghnides continued creating and patenting inventions almost until his death in 1988 aged 87. The Rhino may never have made it past the prototype stage, but we use his water aerator every day; not a bad legacy for a WETech inventor.

The smaller of the Rhino's two prototypes was scrapped by Marmon-Herrington. The other was discovered under a tarpaulin in Ohio by Eugene Pock sometime during the 1960s. Pock exhibited it at Indiana's Mid-America Threshing and Antique Show for many years. Then the Rhino disappeared. Although rumour has it that Pock's son plans on restoring it, he did not respond to requests for an update. As a result, the Rhino's exact location and condition remain a mystery.

19 Cosmonaut Recovery Vehicles

Thunderbird 2 was International Rescue's workhorse. A jet-powered cargo carrier piloted by a marionette on a children's television show, it transported a number of useful vehicles to disaster sites. These included the Mole, which tunnelled through the earth, and *Thunderbird 4*, a submarine.

The Soviet Union's ZiL-4906 was a similarly useful vehicle, the difference being that it actually existed. The 4906, known as the Bluebird because of its colour, was a six-wheeled, amphibious search and rescue truck designed to assist in the recovery of Soviet cosmonauts and their space capsule after they landed in the Russian hinterland.

The Soviet need for a versatile search and rescue vehicle became apparent in 1965 when Voskhod-2 malfunctioned, forcing the crew to land unexpectedly in the wilderness. The two cosmonauts spent the next few days in deep snow, trying to make their way out of the forest on improvised skis before being rescued.

The ZiL-4906, also known as the Bluebird: a six-wheeled, amphibious, search and rescue truck designed to assist in the recovery of Soviet cosmonauts.

Designed by Vitaly Grachev who oversaw ZiL's* specialised design bureau, the Bluebird came in different configurations. One version was equipped with twin cranes to retrieve the Soyuz capsule, which was tied down on the Bluebird's flatbed. Another version was configured as a personnel carrier to transport cosmonauts to safety.

---

* ZiL is an acronym for Zavod imeni Likachyova, a Moscow-based manufacturer of cars, trucks, limousines, military vehicles and heavy equipment. The company went bankrupt in 2012 and was liquidated.

Built during the 1970s, the Bluebird could travel on conventional roads at 50mph. Like *Thunderbird 2*, its job was to get as near as possible to where it was needed. But there were places where even the Bluebird couldn't reach. In those cases, it transported the ZiL-2906 on its back.

The ZiL-2906 was an all-terrain vehicle designed to complete the last mile. Able to reach inaccessible areas where a helicopter or four-wheel-drive vehicle couldn't penetrate, it was propelled by two giant, screw-shaped cylinders, making it look a bit like the Mole.

The ZiL-2906 was able to reach inaccessible areas where a helicopter or four-wheel drive vehicle couldn't penetrate.

The 2906's cab sat atop twin screws, each of which was powered by a separate engine. When it wanted to move forward each screw turned in the opposite direction from the other to counteract lateral forces. When the driver wanted to back up, he used the two levers on either side of his seat to put the screws in reverse. The 2906 also had two four-speed transmissions, one for each screw.

The 2906 steered like a tank. When the driver wanted to turn left, he stepped on the left foot brake, which stopped that screw rotating and allowed the right screw to pull the vehicle in the intended direction. In addition to going forward and reverse, it could also spin in a narrow 360°F circle as well as go sideways.

The 2906 may have been small and light, but it was anything but dainty. Watching a video of it tearing through the woods, it's alarming how aggressive it looks. And because its twin screws were hollow it could ford a lake or river, whether or not it was frozen.

Since the 2906 was first to reach the cosmonauts, it travelled with a doctor. There was also seating in the back for cosmonauts as well as room for two stretchers if needed.

The 2906 was hearty but basic. Two heaters kept its passengers from freezing in the winter. About the only luxury were the front seats, which came equipped with seat warmers. A transponder and radio kept the 2906 in touch with its mother ship.

While the 2906 was versatile, it wasn't without an Achilles heel. It didn't do well with rocks in its path. It also took nearly half an hour to deploy from the Bluebird – precious minutes in below zero weather if a cosmonaut was injured.

What makes the ZiL-2906 a White Elephant is that only five were built. It was so specialised there was very little use for it other than the purpose for which it was designed. Although the Bluebird is still in use today for a variety of applications, the 2906 is retired – not a surprise given it's a WETech vehicle.

An example of the ZiL-4906 and 2906 can be seen at the ZiL Museum in Moscow. There may also be one on Tracy Island, but visiting is prohibited.

20 & 21 Polar Cars and Motorised Sleighs

The Antarctic is one place WETech inventions consistently run into trouble, especially since dog sleds, skis and man-pulled sledges were far superior for crossing the frozen wasteland.

Sir Ernest Shackleton was a polar explorer who had a lot of experience with failure. His Antarctic expeditions are the stuff of legend; his determination, inspirational. But 'the Boss', as his men called him, had trouble getting things right.

On his first attempt to reach the South Pole, called the *Nimrod* Expedition (1907–09), he made the mistake of bringing a motor car. The Arrol-Johnston automobile had been donated by a sponsor to pull sledges filled with supplies. It came with a set of wooden tyres in addition to Dunlop pneumatic ones. Its rear wheels boasted steel studs to improve traction, while its front wheels could be replaced with skis.

The car did OK on the Ross Ice Shelf, but its petrol engine didn't function well in the severe cold. Additionally, the tyres couldn't get traction in deep snow. Unable to traverse the broken ice fields, Shackleton was forced to leave his polar automobile behind.

Shackleton also took a dozen Manchurian ponies on that trip, using them to pull sledges instead of dogs. The ponies proved almost as useless as the automobile. Not only did the expedition

The custom Arrol-Johnston taken on the *Nimrod* Expedition, 1907–09. (Archive of Alfred Wegener Institute for Polar and Marine Research)

fail to reach the South Pole, its ponies didn't make it either. One fell into a crevasse; the rest were shot and eaten.

Captain Robert F. Scott took three motorised sledges to the Antarctic on his *Terra Nova* Expedition (1910–12). Considered the latest in technology, the Wolseley/Hamilton motor sleighs were built to haul sledges filled with supplies across the ice but they were of little use. Their iron caterpillar-type tracks kept coming loose and they were constantly breaking down. As proof they were White Elephant Technology, Scott wrote in his diary, 'I am secretly convinced that we shall not get much help from the motors.'

A Wolseley motor sleigh leaving the 'winter garage' during the *Terra Nova* Expedition. (Kinsey, Joseph James (Sir), 1852-1936: Photographs relating to Antarctica and mountaineering. Ref: PA1-f-067-053-1. Alexander Turnbull Library, Wellington, New Zealand)

Norwegian explorer Roald Amundsen had far more success using skis, dogs and man-pulled sledges, one reason why he beat Shackleton and Scott to the South Pole. But the Boss didn't learn from Scott's experience. On his aptly named *Endurance* Expedition (1914–17), he took motorised sledges to ferry his supplies from one side of the continent to the other. The expedition turned into an epic fail when his ship was crushed in the ice before they'd even started. Shackleton managed to grab victory from the jaws of defeat by safely leading his crew on an incredible journey without losing a single man.

As a result, there is much to admire about the Boss. His family motto, 'Never for me the lowered banner, never the last endeavour', beautifully captures his mindset. Sadly, he died of a massive heart attack at the young age of 47. He was on his way to another polar endeavour when stricken and he was buried on South Georgia Island, the last stop before reaching Antarctica.

William Lashly standing by a Wolseley motor sleigh during the *Terra Nova* Expedition, November 1911.

When the *Nimrod* Expedition was over, Shackleton's car was shipped to New Zealand. What happened to it after it arrived remains a mystery. As for Scott's three motor sleighs, one was lost through the ice and it sank to the bottom of Erebus Bay. The other two sleighs were abandoned after they broke down for the last time. Later, men from the *Aurora* (part of Shackleton's *Endurance* Expedition) found Scott's two motor sledges and repaired them. Once again, they proved useless and were abandoned. As far as anyone knows, they're still out there.

22 Admiral Byrd's Antarctic Snow Cruiser

The Antarctic Snow Cruiser was built to support Rear Admiral Richard E. Byrd's third Antarctic expedition. Designed by Thomas C. Poulter in 1939 and custom-built by Chicago's Pullman Company, the $150,000 vehicle was meant to transport scientists and their equipment on a twelve-month journey to the South Pole and back.

Everything about the Snow Cruiser was big. At 55ft long, it was the largest vehicle ever built at the time. It sported four inflatable Goodyear tyres, each 3ft wide and 10ft tall. It also slept four comfortably, had a machine shop, darkroom, radio shack and galley, and carried a year's supply of food.

Painted fire engine red with an orange and white stripe to be easily spotted in the snow, it could operate at -75°F. On top of everything, it carried a five-passenger Beechcraft biplane on the roof for aerial surveys.

The only way to get the Snow Cruiser from the Chicago factory where it was built to Boston Harbor where it was to be loaded

The Armour Research Foundation published a magazine on the Snow Cruiser Antarctic expedition. (Armour Research Foundation/Illinois Institute of Technology)

A crew member works with a Primus torch to thaw out the Snow Cruiser's wheels on 23 August 1940. (C.C. Shirley/ United States Antarctic Service)

Admiral Richard Byrd, Thomas Poulter and the Snow Cruiser crew, 1939. (Armour Research Foundation/ Illinois Institute of Technology)

aboard a Coast Guard cutter for the journey to the pole was to drive it on surface roads. The 1,000-mile journey took place at 20mph. Thousands of people jammed its route to watch what looked like the world's largest fire truck crawl by.

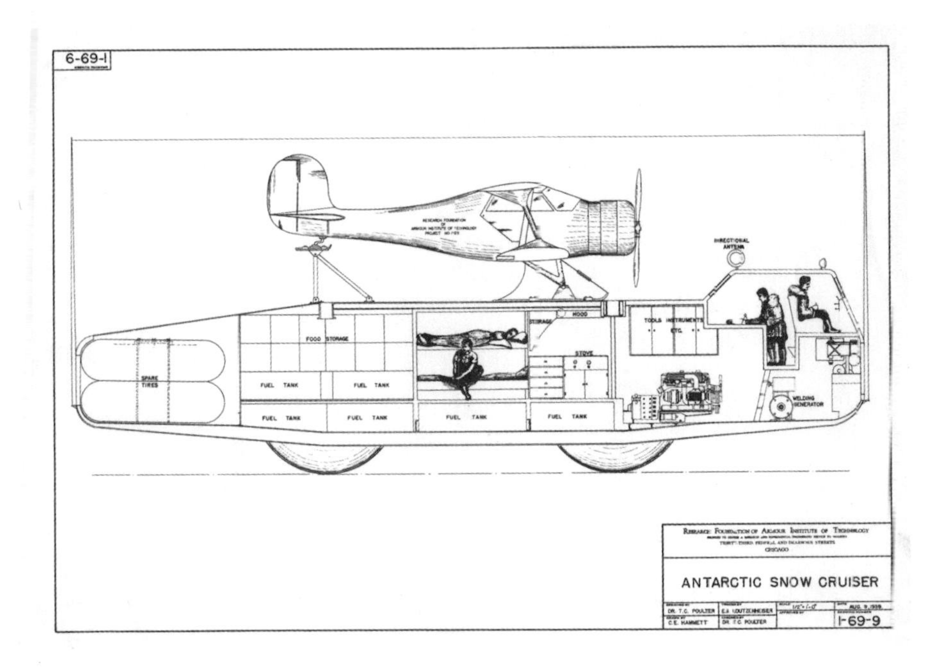

Snow Cruiser diagram. (Armour Research Foundation/ Illinois Institute of Technology)

Byrd's Snow Cruiser ran into trouble the moment it arrived in Antarctica. While being unloaded, the 37-ton behemoth nearly tipped over when it broke through the planks provided to get it ashore. Once on the snow pack it proved ungainly. The cruising speed was limited to 10–13mph, and its excessive weight was so great it constantly sank in the snow. Its nearly treadless tyres made it difficult to get unstuck. Byrd tried attaching chains to the rear tyres, but was unable to overcome the lack of traction. In fact, the best way to drive the Snow Cruiser without getting stuck was backwards. As a result, the longest distance the Snow Cruiser travelled in the Antarctic was 92 miles – and that was in reverse! It covered more ground driving from Chicago to Boston.

The situation must have embarrassed Poulter, who not only worked at the Armour Institute of Technology in Chicago where he designed the Snow Cruiser, but accompanied Byrd on the expedition. As his invention failed to perform in the Antarctic, its nickname, 'the Penguin', seemed sadly suitable, as both performed poorly on land.

The Snow Cruiser was a classic WETech invention because it was expensive, only one was built and it didn't perform as planned. Byrd's expedition ended up abandoning the Penguin in Antarctica and another polar expedition found it in 1946. Allegedly, it only needed servicing to make it operational. Byrd's White Elephant was discovered once again in 1958 with everything exactly as his men had left it, right down to a partially smoked cigarette. Since then, Byrd's Snow Cruiser has disappeared, suggesting it's either buried under a mountain of snow, or more likely sunk to the bottom of the Southern Ocean, a suitable resting place for White Elephant Technology.

23 LeTourneau's Overland Train Mark II

LeTourneau's Sno-Freighter and Sno-Train may have been a good start, but the US Army wanted its 'trackless trains' to carry bigger payloads. The answer was the Overland Train Mark II, designed to cross diverse terrains including deserts and the Arctic. LeTourneau's Land Train was constructed at a cost of $3.7 million. At 565ft long, it is one of the longest vehicles ever built. It had four gas-turbine engines, a six-wheeled locomotive cab, ten cargo trailers (making it look like the Hungry Caterpillar) and two power-generating cars.* Its 30ft-tall command car was so roomy it slept six and came with a galley and toilets.

The Land Train also carried up to 150 tons of cargo at the stately speed of 20mph. This was equal to sixty 2.5-ton trucks, or

* The army briefly considered using a nuclear reactor to power LeTourneau's Land Train, before thinking better of it.

The Overland Train Mark II. (R.G. LeTourneau The Man, Machines, and Mission Collection)

four times the Sno-Train's capacity. When it needed to extend its 400-mile range, more fuel trailers could be added. It even came with radar.

In 1962, LeTourneau shipped his Land Train to the army's proving ground in Yuma, Arizona. Although testing went well, the army cancelled the project shortly after its arrival because heavy-lift helicopters had made it obsolete.

By this point you might ask yourself why so many WETech inventions are funded by the military. Well, the military has the deepest pockets, not to mention the most exotic needs. It's also willing to keep ploughing money into an invention, chasing success long after its clear it's a waste of money. No wonder WETech inventors love the military.

The Land Train's command car, minus its trailers, can still be found at the Yuma Proving Ground Heritage Center in Arizona.

24 Desert-Crossing Vehicles

An ecosystem existed during the 1920s, '30s and '40s for inventors who dreamed big. Venture capital had yet to arrive, but more than half a dozen widely circulated magazines such as *Popular Science*, *Modern Mechanics* and *Science and Invention* were happy to promote an inventor's most exotic musings. At best, many of these inventions existed only as illustrations in a magazine. At worst, they were fantastically impractical with no hope of ever being built.

One class of vehicle that was popular in these pages was desert-crossing vehicles. The French, with their colony in Algeria and protectorates in Morocco and Tunisia, were keen on developing a 'ship to cross the desert', especially one that didn't need to be fed and watered. In 1914, a French lieutenant mounted an enormous six-bladed propeller on the back of a four-wheeled vehicle, hoping to blow himself across the sand at 50mph. The next year, the British mounted a four-bladed propeller on the

The Sizaire-Berwick 'Wind Wagon'.

back of an armoured car with a Vickers machine gun poking over its dash. Called the 'Wind Wagon', it used a rear-mounted aircraft engine tilted downwards for thrust. Since neither vehicle made it past the prototype stage, it's fair to doubt their efficacy.

Getting traction in sand was always a problem for a four-wheeled vehicle. That's why the first motorised vehicle to successfully cross the Sahara was a half-track built by Citroën in 1923. The arduous passage took twenty days and was not for the faint of heart. But that didn't stop an engineer from Kiel, Germany, named Johann Christoph Bischoff from imagining a giant Desert Ship (*Wüstenschiff*) that would cross the Sahara like a luxury ocean liner.

Johann Christoph Bischoff's giant Desert Ship (*Wüstenschiff*) was designed to carry fifty-six passengers across the Sahara like a luxury ocean liner.

Bischoff's creation started out small in 1927. Designed to carry fifty-six passengers in an 86ft-long vehicle, his Desert Ship was to be powered by two diesel engines. But Bischoff's invention soon grew out of all proportion to its original specifications. By 1932, it was a monster-sized vehicle 200ft long and 60ft high. It even looked like an ocean liner with four decks (not counting the pilot house) and an awning-covered promenade for passengers. One article in a German publication said the ship's first-class cabins would offer all the comforts of a luxury ocean liner. They would even be artificially cooled to combat the desert sun.

Bischoff claimed his Desert Ship would carry 300 people and 20 tons of freight across the desert at 30mph. Powered by a series of huge 900hp Maybach engines, it would have a range of 6,000 miles. And since the vehicle would carry enough food, water and fuel to travel for months, crossing the Gobi or Sahara deserts would be a breeze.

One of the more incredible aspects of Bischoff's invention was its four massive wheels, each made of solid iron with a diameter of 50ft. Shaped like the paddlewheels on a Mississippi riverboat, they were designed to provide traction in the soft and shifting sands.

Bischoff built a scale model of his behemoth, which was featured on the cover of *Modern Mechanics and Inventions* in 1931. The accompanying article said construction was 'soon to begin'.

Bischoff viewed his creation as an alternative to camels, roads and trains. It would cross the desert like a ship crosses the sea. But the desert is not the ocean. Torsion (i.e. twisting) would have torn Bischoff's invention apart given the desert's uneven surface. As it was, financing was probably a bigger problem given Germany's runaway inflation.

What's most notable about Bischoff's Desert Ship is that it's an excellent example of the Second Rule of Failure: a project's likelihood of failing increases in direct proportion to its size and complexity. We've learned a lot about building desert-crossing vehicles since Bischoff's time. Just witness the robust performance of NASA's rovers on Mars.

Sadly, the historical record makes no further mention of Bischoff and his ship of the desert, which is not surprising since WETech inventors and their inventions often disappear without a trace.

Chapter Five

All Things Atomic!

> Failure is the tuition you pay for success.
>
> Fortune Cookie

President Eisenhower's 'Atoms for Peace' programme, launched in 1953 to promote the peaceful use of atomic energy, inspired a lot of hare-brained schemes. The optimistic if premature application of a new technology happens all the time. It only makes sense until enough failures accumulate, causing it to slow.

Atomic energy seemed to solve many important problems related to agriculture and medicine in the 1950s, but none more so than transportation. That's why so many atomic-powered trains, planes, cars, trucks, rockets ships, ocean liners and even dirigibles were designed. Thankfully most never got off the drawing board.

For example, the Ford Motor Company unveiled the Nucleon in 1957, an atomic-powered car with a nuclear reactor in its boot. The car was a throwback to the early days of automobiles that used steam for propulsion. In this case, the water wasn't heated by kerosene, but by nuclear fission. Ford claimed the Nucleon would go for 5,000 miles before its uranium needed to be replaced – not very far, considering. Since no reactor existed small or light enough to fit a car, the Nucleon remained a scale model.

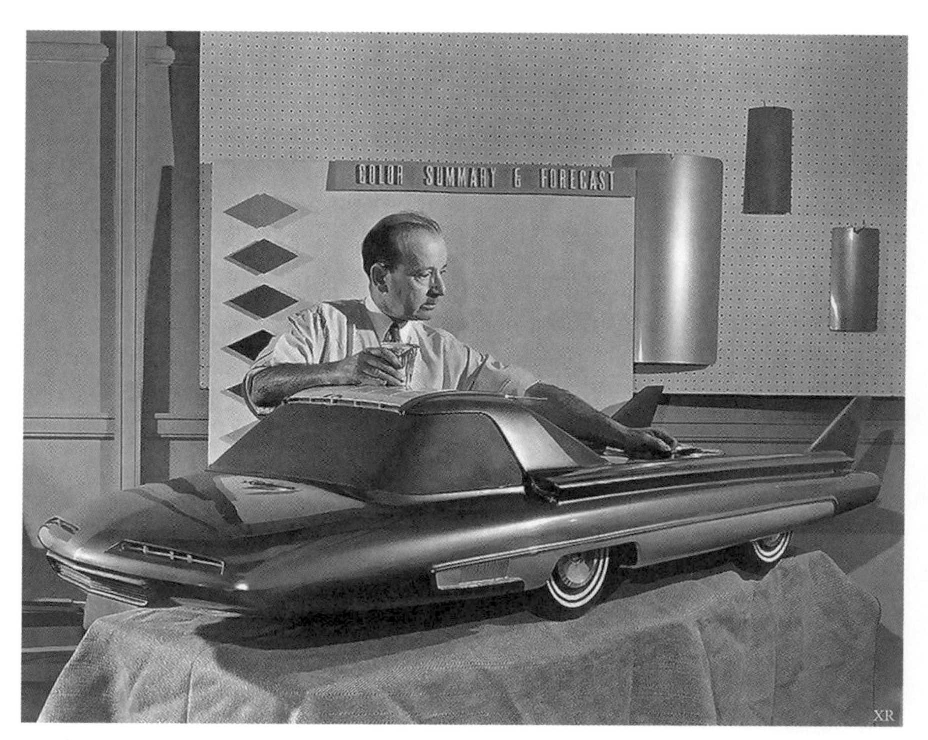

A scale model of
the Ford Nucleon.

Atoms
for Peace
promotional
van.

The Studebaker-Packard Corporation followed suit the next year when it debuted a full-sized concept car, the Astral. Said to be powered by either nuclear energy or 'an ionic engine', the car would use a 'protective curtain of energy' to shield its passengers from radiation. Yeah, right. It also had only one wheel. Designed by Edward E. Herrmann and looking like something out of *The Jetsons*, the only place the Astral went was into storage.

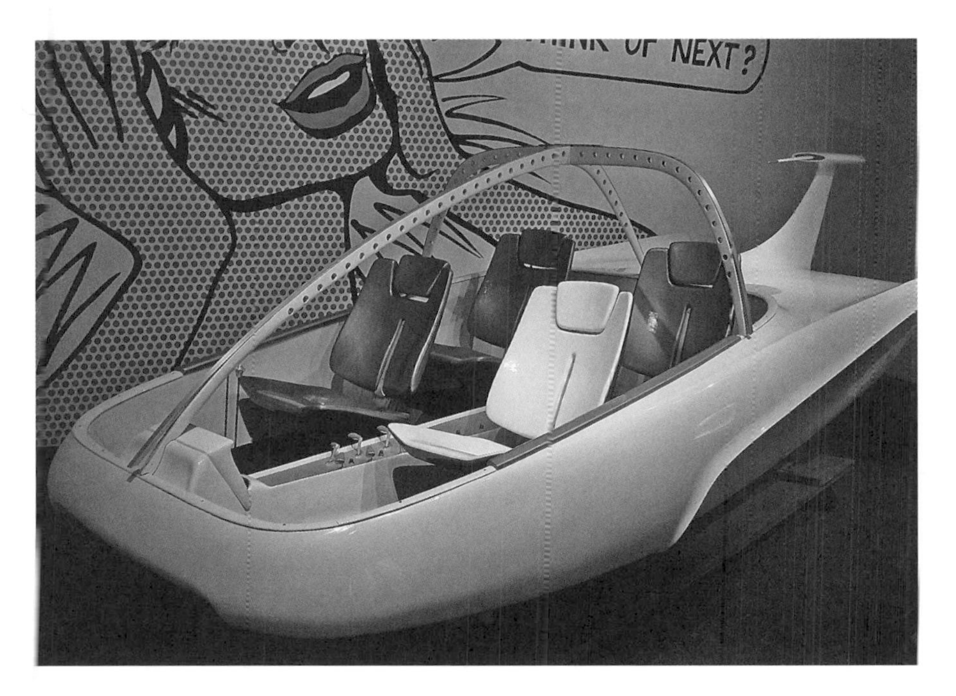

The Studebaker Astral designed by Edward E. Herrmann. (Creative Commons CC BY-NC-ND-2.0)

Cars weren't the only vehicles thought suitable for atomic propulsion. Railways also sought nuclear power. In 1948, five years before Eisenhower's 'Atoms for Peace' initiative, an industry lobby group called the Federation for Railway Progress ran an ad in *Time* magazine asking the question: 'Will your railroad have a place at the atomic research table?' The Federation's answer was predictable: 'No industry stands to benefit more from atomic "vitamins" in its diet than the undernourished railroads.'

Dr Lyle B. Borst, a physics professor at the University of Utah, agreed. With help from his graduate students, he designed the X-12 nuclear-powered train in 1954. The two-unit engine weighed twice as much as a conventional locomotive, and its reactor shielding alone weighed 200 tons. Powered by a steam-generating 'atomic pile', the X-12 was patented by Dr Borst, who'd once designed reactors for the Atomic Energy Commission. But many aspects of Borst's design were unfeasible. For example, its steam-driven turbines would become so contaminated with radiation they could never be touched. Additionally, the cost of building the X-12 was four times that of a regular locomotive. Despite *Life* magazine calling it a 'practical dream',* Borst's atomic train derailed of its own accord.

That didn't stop an industry trade group called the Association of American Railroads from hiring a nuclear physicist to design an atomic-powered locomotive.** The trade association was so confident that nuclear-powered trains were right around the corner it ran an ad in 1960 claiming, 'Railroads offer the greatest opportunities for the efficient use of nuclear energy.' Meanwhile, Republican Senator John Butler wanted to pass legislation authorising the Atomic Energy Commission to develop a nuclear reactor to power locomotives. Butler hoped this would lead to the 'Freedom Train', an atomic-powered train that would tour the United States promoting the peaceful uses of nuclear energy. That never happened. Instead, the 'Atoms for Peace' programme settled on a promotional bus with an internal combustion engine.

Sometimes you can't keep a bad idea down, which is why the Soviet Union's Ministry of Transport made atomic-powered trains a goal during the 1950s. Though nothing came of it, Russian Railways announced as recently as 2011 that it was working with the State Atomic Energy Corporation to design a nuclear-powered locomotive using a fast breeder reactor. Since then, nothing's been heard.

---

* As an interesting side note, Clare Boothe Luce, whose husband was the publisher of *Time* and *Life*, was listed as an adviser to the Federation for Railway Progress, which may explain why atomic-powered trains received such positive coverage in those publications.

** The physicist was hired from the Armour Research Foundation, the same people who built Byrd's Snow Cruiser.

The Atomic Train. (Courtesy American Association of Railroads)

So far none of the efforts to build an atomic train have amounted to much. This is just as well. As Edward Teller, the father of the hydrogen bomb, once said, atomic trains are an ingenious way to 'combine minimum utility with maximum danger'; not that that ever stopped a WETech inventor.

ooo

A 1955 illustration by Frank Tinsley of an atomic-powered airship, promoting President Eisenhower's 'Atoms for Peace' programme.

You'd think an atomic-powered airship was a particularly bad idea given their history of disasters, but a number of proposals for atomic-powered airships circulated during the 1950s and '60s, some more credible than others.*

The first study of an atomic-powered airship was conducted in 1954 by the US military. Designed to carry radar to guard against a Soviet first strike, the airship was considered a more viable platform for nuclear propulsion since it required less power to operate than an aeroplane. Although the report concluded further study was necessary, it suggested many of the problems were solvable.

* If you really want to scare yourself, look up Project Orion, a proposal to detonate 800 atomic bombs to propel a 'starship' in space, or Project Plowshare, which proposed using 520 nuclear explosions to excavate a canal.

The Goodyear Tire & Rubber Company had extensive experience building blimps for the US Navy. It also had a subsidiary that operated atomic reactors. Combine them and you have Goodyear's 1959 proposal to build two types of nuclear-powered airships: one for the military, the other a cargo carrier for the commercial market. Both airships would be operated by a twenty-four-man crew and come with atomic-powered turboprop engines, giving it 'unlimited range'. Despite the hype, neither airship got built.

Another atomic-powered dirigible that dominated discussion came from Francis Morse. Morse was a former Goodyear engineer turned assistant professor of aeronautics at Boston University. His atomic airship generated a lot of publicity during the 1960s, including write-ups in *Time* magazine and *Aviation Week & Space Technology*. Morse admitted his airship presented a serious public safety hazard since a crash could 'spread fissionable material with lamentable consequences'. But he also claimed an airship's inherent buoyancy reduced this danger to a manageable level. Still, financing was as scarce for nuclear-powered airships as it was for nuclear trains, so Morse's proposal died on the vine.

Since none of the above projects advanced past the drawing board, they were pipe dreams. However, more than a few atom-dependent inventions got at least as far as the prototype stage and a few a good deal further. Cold War competition was responsible for some of the crazier ideas listed below.

A model of Ford's Nucleon can be seen at the Henry Ford Museum in Dearborn, Michigan; the Astral can be seen at the Studebaker National Museum in South Bend, Indiana.

25 Atomic-Powered Aeroplanes

Both the United States and the Soviet Union invested heavily in developing an atomic-powered bomber during the 1950s, despite the question of what would happen to all that radioactive material should one of the planes crash or be shot down.

The NB-36H (NB stood for Nuclear Bomber) was a B-36H bomber modified by Convair for the US Air Force. Nicknamed 'the Crusader', it made more than forty-five test flights between 1955 and 1957 with a nuclear reactor on board. The reactor wasn't used to power the aircraft, however. Rather it was used to measure the effectiveness of the crew compartment's radiation shielding. Meanwhile, both Pratt & Whitney and General Electric were hard at work developing a nuclear-powered turbojet engine to power the Crusader's successor, the Convair X-6.

The X-6 was to be a fully operational nuclear-powered bomber, but President Eisenhower was never convinced of the billion dollar programme's utility. When President Kennedy took office in 1961, he cancelled the project because intercontinental ballistic missiles were quickly rendering a nuclear-powered bomber obsolete.

NB-36H – the Crusader atomic aeroplane.

26 Atomic Lighthouses

Using atomic energy to power a lighthouse seems an expensive solution compared to paying two humans for their upkeep, but it made sense to the Soviet Union for more than fifty years.

The USSR first began building lighthouses along its remote northern coast to guide merchant ships during the 1930s. Since then, 132 nuclear-powered lighthouses have been built in places where neither people nor electricity was present. Powered by a nuclear battery called a radioisotope thermoelectric generator (RTG),* the Soviet lighthouses were fully autonomous and dispensed with humans altogether.

The use of radioisotopes to generate electricity was common in the Soviet Union. There was even a slogan 'Nuclear energy in every house!' A Soviet RTG used strontium-90 (Sr-90) to produce heat, which a battery then converted into electricity to power the lighthouse lamp. With a half-life of only thirty years, Sr-90 was thought ideal for powering lighthouses in remote areas. But strontium-90 is the waste product of a nuclear reactor. It's not only highly radioactive, it's extremely toxic. It's also an important ingredient in making a dirty bomb.

The Soviet Union's atomic-powered lighthouses remained in operation until the 1980s. But once the USSR collapsed, maintenance lapsed for more than a decade. Several RTGs disappeared during this time, either swept into the sea or stolen by looters. In 2001, two people at a Soviet lighthouse north of the Arctic Circle were exposed to radiation when they tried prying loose the RTG's lead shielding to sell as scrap metal.

Decommissioning the nuclear batteries finally began in 1996. A considerable job, it took some time to accomplish. All RTGs have since been removed from Russian lighthouses, but at least two remain missing. Yet another reason the project is a poster child for White Elephant Technology.

* The RTGs used in Soviet-era lighthouses were named Beta-M.

It's possible to visit a Soviet lighthouse once powered by atomic energy but they're probably best avoided.

27 The Atomic Tank

Although Eisenhower's 'Atoms for Peace' programme focused on the humanitarian use of nuclear energy, the military showed plenty of interest.

During the 1950s, Chrysler considered using nuclear power to propel an experimental tank it hoped to sell to the US Army. A nuclear-powered, vapour-cycle plant was deemed one of several options for Chrysler's TV-8 tank because it eliminated the need for constant refuelling.

A full-size prototype of Chrysler's TV-8 tank.

Armed with a 90mm gun, two .30-calibre machine guns and a remote-controlled .50-calibre machine gun on its turret, the TV-8 was to be equipped with video cameras to relay outside images to the television screens inside the tank. If this wasn't Buck Rogers enough for 1955, the TV-8 was also intended to be amphibious, enabling it to attack ships at sea.

You'd think the military would have left well enough alone after its failures with tanks that flew, swam or submerged, but this wasn't the case. The problem with a nuclear-powered tank is that it's a rolling atomic bomb capable of destroying everything around it if the enemy hits it.

There's not much in the public record about the TV-8. What qualifies it as White Elephant Technology is not just that Chrysler considered using nuclear power to propel it, but that they actually built a full-scale model. Eventually, saner heads prevailed and the project was dropped before a fully functional prototype was built.

28 The NS *Savannah*

Senator Butler may never have got his atomic-powered 'Freedom Train', but the Eisenhower administration did build the first atomic-powered merchant ship, the NS *Savannah*, in 1959. Constructed at a cost of $47 million, more than half of which went on her nuclear reactor, the NS *Savannah* (NS stands for Nuclear Ship) was intended to showcase the peaceful uses of atomic energy.

The respected naval architecture firm George G. Sharp, Inc. designed the ship, which was built in a New York City shipyard. Her reactor was provided by Babcock & Wilcox, experts in building nuclear power plants.

The Eisenhower administration made a big deal of the *Savannah* since it was the world's first atomic-powered passenger/cargo ship. Vice President Richard Nixon's wife, Pat, presided over the keel-laying ceremony. Waving an 'atomic wand', she triggered a Geiger counter that signalled a crane operator to lay down the keel's first section. President Eisenhower also lent his wife to christen the ship in 1959, confirming its importance.

The *Savannah* was two football fields long and weighed 14,000 gross registered tons. It also had a sleek, futuristic appearance that included atom graphics on the hull and an interior design with a modern, Atomic Age look.

The control room of NS *Savannah* MD1.

But the ship was plagued by a series of mechanical problems as well as a painful labour strike that drew negative media attention. Despite giving 1.4 million tours and travelling half a million miles to visit forty-five nations and thirty-two US ports, the *Savannah* was prevented from docking in Japan, Australia and New Zealand due to their no nukes policy.

High operating costs eventually doomed the ship. After seeing only ten years of service, it was decommissioned in 1971.* Despite its short life, the *Savannah* appears in the National

* The Soviet Union has had more success with nuclear-powered commercial ships in part because they are subsidised. These include six ice breakers and at least one cargo ship.

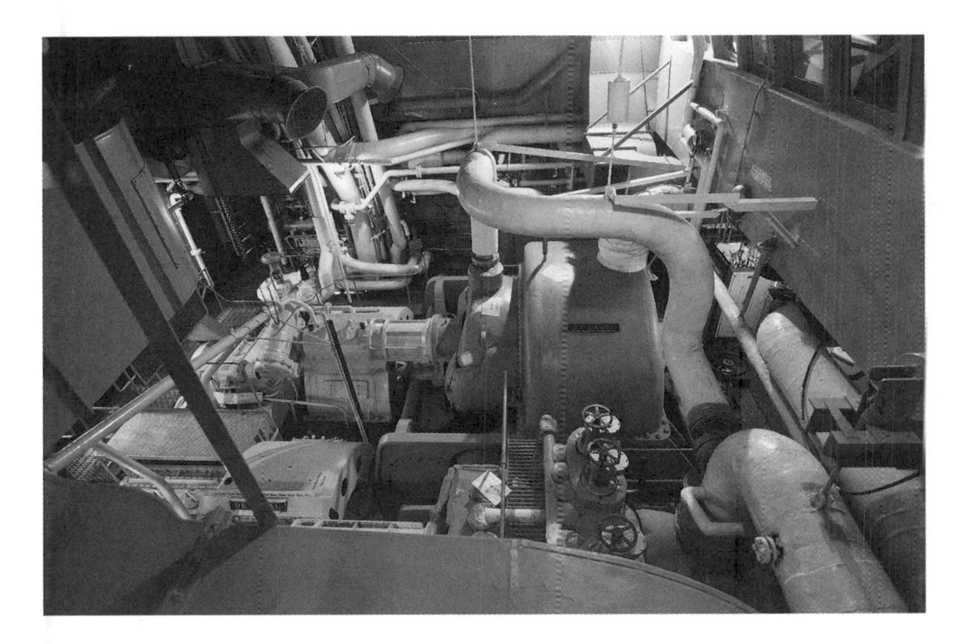

NS *Savannah*'s engine room.

Register of Historic Places and has been designated a National Historic Landmark. Since only one was built, it also makes our list of fifty crazy WETech inventions.

The NS *Savannah* is currently moored at Pier 13 at the Canton Marine Terminal in Baltimore, Maryland.

29 The Mini Atomic Crawler

A shipboard nuclear reactor is expected to travel from place to place, but what about a nuclear reactor on tank treads?

The TES-3 (Transportable Electric Station)** was the Soviet Union's first experiment in building a mobile nuclear reactor. Intended to provide power for mission critical facilities such as hospitals in disaster areas, the TES-3 began generating

** Also known as Object 27.

electricity in 1961. Despite its altruistic cover, the TES-3's primary purpose was to provide power to the Soviet military's remote missile and submarine bases.

The TES-3's reactor and support equipment were mounted on top of four, self-propelled, caterpillar-tracked bodies based on the Soviets' T-10 heavy tank chassis. The entire thing weighed 310 tons, 37lb of it uranium. Tests were conducted throughout the 1960s, but the TES-3 only generated 1.5 megawatts of power, less than a modern wind generator. When the project was deigned unprofitable it was cancelled in 1969, which is surprising since a lack of profitability has never stopped the military. More likely, the Soviets realised that a radioactive power plant on tank treads was impractical.

If you think that put an end to mobile nuclear reactors you're wrong. In the 1980s, the Belorussian Academy of Science built a 60-ton truck designed to carry a reactor. Called the Pamir, it consisted of four trailers: one for the reactor, another for the gas turbine, one for the control room and the last housing the twenty-eight staff necessary to operate it all. The project was completed in 1986, just in time for Chernobyl. The project was cancelled shortly thereafter.

But that wasn't the end of mobile atomic reactors. In 2020, Russia commissioned the world's first floating nuclear power

The Soviet TES-3 Mini Atomic Crawler, a nuclear power plant mounted on self-propelled tracked vehicles.

plant. Built by the State Atomic Energy Corporation (the same people who wanted to build an atomic-powered train), the floating plant now delivers power to the electrical grid on the Siberian coast.

30 The Atomic Cannon

If the 'Atoms for Peace' programme was meant to show nuclear energy could be used for things other than Armageddon, the atomic cannon demonstrated the opposite. Nicknamed 'Atomic Annie' but officially known as the M65 280mm Motorized Heavy Gun, it was a giant artillery piece designed to fire a nuclear warhead. Developed by the US Army during the early 1950s, the atomic cannon was said to be based on the German Army's railway gun 'Anzio Annie', which was used during the Second World War to defend Italy against the Allies.

A huge artillery piece weighing 47 tons isn't much good if you can't get it somewhere, so two trucks, one for the front of the atomic cannon and the other for the rear, were designed to tow it. Built by the Kenworth Motor Truck Company, the trucks looked like a cross between a hook and ladder fire engine and Dr Dolittle's Pushmi-Pullyu.

Atomic Annie's moment of truth came at the Nevada Proving Ground in 1953. That's when soldiers carefully loaded an artillery shell with a W-9 warhead into the breech. According to film footage, the M65 demonstrated an impressive recoil when fired. It took nineteen seconds for the 15-kiloton warhead to travel 7 miles before detonating above ground. Among the objects it destroyed was a yellow school bus; an odd choice given her primary target was Soviet tanks.

The maximum range of Atomic Annie was officially listed as 18 to 20 miles, although at least one report suggests it was closer to 35. Twenty atomic cannons were manufactured at a cost of $800,000 each. The majority were deployed in West Germany, followed by Okinawa in Japan, and South Korea.

The M65 280mm Motorized Heavy Gun, nicknamed 'Atomic Annie', being tested at the Nevada Proving Ground.

The M65 was important enough that the US Secretary of Defense was present for its test firing. Fortunately, it was the first and last time a cannon ever fired a nuclear device. The truth is that the army's tactical nuclear missiles had already rendered Atomic Annie obsolete by the time it was deployed, but since it was considered a 'prestige weapon' it wasn't retired until 1963.

The only M65 280mm Motorized Heavy Gun to fire a nuclear shell can be seen at the US Army Artillery Museum at Fort Sill in Oklahoma. Other M65s can be found at the Armoured Fighting Vehicles Museum at the US Army's Aberdeen Proving Ground in Aberdeen, Maryland; the National Museum of Nuclear Science and History in Albuquerque, New Mexico; Freedom Park in Junction City, Kansas; the Rock Island Arsenal in Rock Island, Illinois; the Watervliet Arsenal Museum in Watervliet, New York; and the Yuma Proving Ground at Yuma, Arizona.

31 The Davy Crockett Nuclear 'Mortar'

If a mobile nuclear cannon doesn't worry you, maybe a portable, field-launched 'mortar' firing a nuclear projectile will.

The M28 and M29, also known as the Davy Crockett Weapon System, was one of the smallest nuclear weapons ever built. Taking command and control of a nuke out of the hands of the president, it put it squarely in the hands of three infantry grunts.

Tactical nuclear weapons not only provided more bang for the buck, they were key to countering the Soviets' huge advantage in ground forces. As a result, Army Chief of Staff General Maxwell D. Taylor considered tactile nuclear weapons a top priority. After investigating more than twenty potential delivery systems, the army settled on a recoilless rifle because it was the simplest, lightest option to deploy at the front lines.

The main difference between the M28 and the M29 was weight and range. The M28 weighed 185lb and had a range of 1.25 miles, while the M29 weighed 440lb with a slightly longer range of 2.5 miles. Both variants fired the M388 projectile with a W54 warhead – the smallest nuclear weapon in the US arsenal. Since the nuclear projectile was oval-shaped, soldiers called it 'the atomic watermelon'.

The yield of the W54 warhead was 10 to 20 tons of TNT, considerably less than that of *Fat Man* or *Little Boy* dropped on Japan during the Second World War. This was the point since the much smaller explosion was meant to be 'tactical' rather than 'strategic'.

The Davy Crockett was operated by a three-man crew. The lighter version could be launched from a jeep or from the ground using a tripod. The heavier version was usually transported in an armoured personnel carrier.

After the three-man crew fired a spotting round to determine trajectory, the M388 projectile was loaded into the launcher's barrel like a rifle grenade. Once fired, four fins deployed on the projectile's tail, stabilising it in flight. Although there are

The Davy Crockett Nuclear 'Mortar'. (US Army)

conflicting reports, accuracy appears to have been a problem. Then again, how important is accuracy when it comes to a nuclear explosion?

A combination of heat, concussive force and neutron radiation would have proved lethal to any enemy within the blast radius. However, the weapon's short range also made it a threat to its three-man crew. No wonder the army urged them to seek shelter immediately after launching the device.

The army began deploying the Davy Crockett Weapon System in 1961. Borrowing from the lyrics of the old theme song, 'Davy, Davy Crockett, king of the wild frontier', the wild frontier during the Cold War was the border between East and West Germany. This is where the Davy Crockett System was deployed as well as in Guam, Hawaii, Okinawa and South Korea. More than 2,000 nuclear mortars were manufactured. Fortunately, none saw combat. The only one to fire an atomic projectile did so as part of a training exercise designed for soldiers to experience the nuclear battlefield.

The Davy Crockett's service was relatively brief, lasting only ten years. By 1971, it had been retired, not just because of its escalating cost, but what Brigadier General Alvin Cowan said was the 'great fear that some sergeant would start a nuclear war'.* Don't breathe a sigh of relief, though. Russia, China, France and the United States still have plenty of tactical nukes.

Davy Crockett systems can be found at the Don F. Pratt Museum at Fort Campbell, Kentucky; the National Museum of Nuclear Science and History in Albuquerque, New Mexico; the West Point Museum at West Point, New York; the National Museum of the United States Army, Fort Belvoir, Fairfax County, Virginia; the Air Force Space and Missile Museum, Cape Canaveral Space Force Station, Florida; the National Infantry Museum, Fort Benning, Georgia; Watervliet Arsenal Museum, Watervliet, New York; the National Atomic Testing Museum, Las Vegas, Nevada; and the Don F. Pratt Memorial Museum at Fort Campbell in Clarksville, Tennessee.

* From 'Proceedings of Tactical Nuclear Weapons' by the United States Atomic Energy Commission and Department of Defense, September 1969, p.173.

Chapter Six

Failures vs Frauds

> It is hard to fail, but it is worse never to have tried to succeed.
>
> Teddy Roosevelt

History is loaded with hucksters who promoted a fake invention only to raise money they later absconded with. Still, it can be difficult to tell the difference between an invention that falls short of expectations and one that is an outright fraud. This is especially true given how often inventors overpromise what their invention can do. Hence, in the following survey you can decide which inventions are failures and which are frauds.

32 Dr Moller and the Moller Skycar

Canadian inventor Dr Paul Moller has spent nearly sixty years and more than US$100 million perfecting flying cars that have never flown unassisted.*

One of his earliest iterations was the Discojet, a hovercraft that looked like a flying saucer with duct fans and a bubble canopy. A prototype flew briefly while tethered to a crane in 1965, but it only hovered 2ft off the ground before losing stability. Since then improved versions have made test flights, but none very high, or for very long.**

Like a lot of WETech inventors, Dr Moller hoped to sell his invention to the military, but they didn't bite. Neither did the commercial market. That didn't stop Dr Moller. In 1983, he founded Moller International (MI), not only to market the Discojet, which he'd rebranded the Moller 200X, but to design, develop, manufacture and market the company's breakthrough invention: the Moller M400 Skycar.

The Skycar was an aesthetic and technological improvement over the Discojet. Combining the characteristics of a fixed-wing aircraft with the vertical take-off and landing (VTOL) capability of a helicopter, the Moller Skycar was ahead of its time.

The real breakthrough was its eight Wankel rotary engines licensed from a German company. Generating a top speed of over 350mph, they provided the power-to-weight ratio necessary for vertical take-off. They were also inexpensive to produce. At one point, Dr Moller estimated annual Skycar sales would reach several thousand units, starting in 1995. It didn't work out that way.

That year came and went without any sales for the bearded, blue-eyed inventor with a penchant for oversized glasses, but 2003 turned out to be his watershed year. In a proof-of-concept

* Some reports suggest Dr Moller may have spent up to $200 million.

** The M200X's most impressive test flight was in 1985, when it hovered at 60ft for two minutes.

test flight, the four-seat Skycar hovered off the ground while tethered to a crane. That same year, Dr Moller raised $5.1 million in a stock offering. He also began accepting deposits on the Skycar, just like Elon Musk did before he launched the Tesla.

Unfortunately, the Securities and Exchange Commission (SEC) charged Moller with civil fraud, claiming he'd sold unregistered stock to the public over the internet. He was also charged with inflating sale projections to drive up his company's stock price. Dr Moller settled the suit without declaring guilt, paying a $50,000 fine, but the negative publicity put a chill on financing.***

Work on Moller's Skycar slowed after that. There were sporadic announcements related to the project, but not much seemed to be going on. Several years after settling with the SEC, Dr Moller tried auctioning the Skycar's only working prototype on eBay. When no one met his $3.5 million reserve, the minimum bid was lowered. When the Skycar still didn't sell, the auction was cancelled.

In 2007, a toy company licensed the rights to manufacture a die-cast replica of the Moller Skycar, which it still sells today. That same year, Dr Moller announced the M200G Volantor, a rebranded version of his Discojet with upgraded technology and a two-person cabin.****

But things did not go well for Dr Moller. By 2009, Moller International was in debt to the tune of $40 million. That's when he filed for personal bankruptcy. That same year, the M400 Skycar appeared in the Neiman Marcus holiday catalogue with a price tag of $3.5 million, but it didn't sell.

In 2013, Dr Moller launched a crowd-funding campaign for the Skycar, but raised only 3 per cent of his goal. After that, Moller International went dormant and its website shut down (never a good sign).

*** Securities and Exchange Commission; Litigation Release No. 17987, 19 February 2003; SEC vs. Moller International, Inc., and Paul S. Moller, Defendants (USDC, Eastern District of California, Sacramento Division, Civil Action No. 2:03-CV-261 (WBS) DAD.

****The Discojet has been rebranded several times. Names have included: the Discojet, the M200X, the M200G, the M200 Volantor, the Skycar 200 and the Neuera.

Rumours circulated that Dr Moller was negotiating a deal to open a factory in China. 'I'm out grubbing for money everywhere rather than doing the technical things,' the inventor told *Sactown* magazine. 'Money, it's always money.'

In 2017, a news story reported that Dr Moller was again trying to sell his Skycar prototype. He also released his last newsletter in that year. In it he said Moller International 'is not interested in promoting the Skycar further. It only brings additional interest … from the general public, and they … question why MI has not yet produced it.'

ooo

The M400 Skycar. (Eric Paul Pierre Pasquier/Getty Images)

Paul Moller and his M400 Skycar. (San Francisco Chronicle/ Hearst Newspapers/ Getty images)

It's important to understand that Dr Moller is no crackpot. He holds a Master's degree in engineering and a Ph.D. in aerodynamics from McGill University. He was also a professor of mechanical and aeronautical engineering at the University of California, Davis, from 1963 to 1975, where he developed its aeronautical engineering programme. Versions of his Skycar have appeared on the cover of *Popular Science*, *Popular Mechanics* and *Forbes FYI*. In 2003, *Esquire* magazine gave him its Genius Award.

Dr Moller first became interested in VTOL aircraft in 1957 when he was a 20-year-old summer hire working with the Canadian Defense Research Board. There he saw plans for Project 1794 – a flying saucer-shaped VTOL aircraft being developed by Avro for the US military.

'I had a security clearance, so I got to see and learn a lot about that particular project,' Moller told *Sactown* magazine. It's also what inspired him to build the Skycar.

ooo

Things continued heading south for Dr Moller. In 2019, the SEC delisted shares of Moller International from the over-the-counter market. That same year, the *Mercury News* reported two of Dr Moller's prototypes had been destroyed when the barn on his property burned down in a fire.

Incredibly, none of these setbacks have stopped Dr Moller pursuing his dream. Moller International now operates out of Dixon, California, where Dr Moller lives. According to the company website, MI remains determined 'to develop and put into use personal transport vehicles that are as safe, efficient, affordable, and easy-to-use as automobiles'. The website includes grainy film clips of previous test flights, some more than forty years old. They're joined by a computer rendering showing what the Skycar, now in its 'fifth generation', will look like. There's no mention of a fire destroying any prototypes.

The maroon-coloured Skycar certainly wins first prize for 'best-looking' flying car, but it appears no closer to mass production today than forty years ago. Eighty-six years old and thrice married, time is running out for Dr Moller. He remains *persona non grata* among VTOL investors. Although the 'flying car' business is blossoming with many new competitors, rivals feel Dr Moller's travails taint the industry. Some critics call the Skycar vapourware, while others point to Dr Moller's stock sale as an act of bad faith.

So, is Dr Moller a failure or a fraud? Well, it depends on whom you talk to. Many Skycar investors are unhappy, but what entrepreneur isn't guilty of optimistic sales projections? And what inventor doesn't run short of funds they need to make their invention a success? Does that make the Skycar a fraud? Probably not. Instead, Dr Moller is a good example of the Third Rule of Failure: inventing always takes longer and costs more than planned.

Whether Dr Moller is an underfunded visionary ahead of his time or a huckster,* one thing is clear: he truly believed in the Skycar and was willing to do almost anything to ensure its success. Yes, he paid a steep price for chasing his dream, but there are far worse things a man can be guilty of.

33 The Cooley Air-Ship

Many inventions fail to live up to their hype, but does that mean an inventor is out to defraud his investors?

When John F. Cooley promised in 1910 to build the world's first aeroplane capable of crossing the Atlantic, the furthest an aeroplane could fly was 110 miles.** Calling his invention the Cooley Air-Ship, Cooley's intentions were unclear even after he disappeared with investors' money. Was he a genuine inventor or a conman? The evidence points both ways.

Cooley formed the Cooley Aerial Navigation Company in 1910 to design and build the largest aircraft that had ever flown.

---

* Dr Moller did not respond to repeated requests for an interview.

** The Wright Model B.

Nicknamed 'The Dreadnaught of the Skies', Cooley's Air-Ship was 81ft long, weighed nearly 3,000lb and came with four wings. Designed to accommodate a pilot, flight engineer and two passengers, the canard-style aircraft looked unlike anything that had ever flown.

Initial work on Cooley's Air-Ship began in New York City. Later, he moved operations to a farm near Rochester, where he built his aeroplane in a tent. As the Air-Ship took shape it looked like something out of a Jules Verne fantasy. The cabin was enclosed to protect passengers from the elements, an innovation at the time, but was ridiculously cramped at less than 3ft wide. Boasting large, porthole-style windows and a periscope for the pilot, the plane was constructed from spruce wood, cloth and piano wire.

Cooley's design called for two 90hp engines to power his plane at speeds up to 125mph. For perspective, the Wright Model B's engine was only 42hp and flew at an average speed of 44mph. Cooley also described a pneumatic control system similar to hydraulics. Using carbonated gas in an on-board tank to maintain pressure, it was designed to assist the pilot in handling the flaps on the wings and tail. With a proposed range of 300 to 500 miles, Cooley's Air-Ship was way ahead of its time. But it had to fly first and that took money, lots of it.

Cooley initially told people his Air-Ship was backed by a syndicate of wealthy New Yorkers. Later, he held fundraising dinners at a Rochester hotel to solicit money to complete his aircraft. The highlight of the banquet came after dessert when Cooley, who sported a moustache the ends of which looked like broom bristles, flew a scale model of his Air-Ship in the banquet hall to everyone's delight.

On 4 December 1910, the *Rochester Herald* announced the Cooley 'airship' was nearing completion. Cooley had planned on flying his creation from Rochester to New York, but at some point changed his mind to crossing the Atlantic – an obvious impossibility given the plane's purported range.

Cooley estimated the cost of building his Air-Ship would be $12,000 (an amount he'd already exceeded by the time of the *Herald*'s announcement). In fact, Cooley had taken in at least

$20,000 without flying anything more than a scale model. Construction continued throughout the winter as rumours flew. Come spring, Cooley had vamoosed with the dough.

The Cooley Air-Ship. (George Grantham Bain Collection/ Library of Congress)

Reports suggest Cooley was in New York raising funds via a stock sale, but when he failed to return to Rochester, people accused him of being a flimflam man.

Since Cooley's workers were no longer being paid, construction on his Air-Ship came to a halt. Next, his unpaid grocery bill in the amount of $92 resulted in the local court issuing a legal decision attaching his plane to pay off his debts. A lawyer involved in the case doubted anyone would buy Cooley's Air-Ship. 'It may be worth … thirty cents, it may be worth a million,' the lawyer said. Nobody knew its true value. Meanwhile, Cooley's folly was left to rot on the banks of the Genesee River.

Many believe Cooley to be a scam artist, but a closer examination calls that into question. For thirteen years, including the time he spent working on his Air-Ship, Cooley filed nearly twenty patents related to his aircraft. If he was a conman, why would he go to the trouble of filing patents over such a long period? A more

likely explanation is that Cooley was sincere if misguided. After all, it was the early days of aviation when many thought they could improve upon the art. As expenses exceeded income and he was no closer to flight, Cooley must have felt so overwhelmed he took flight himself. It wouldn't have been the first time such a thing had happened.

Cooley with his Air-Ship. (Rochester Museum & Science Center)

Passengers in the Cooley Air-Ship. (George Grantham Bain Collection/ Library of Congress)

Or maybe Cooley engineered the whole thing just to get the money. No one knows for sure. Which means the Cooley Air-Ship may be lost to time, but remains a mystery.

34 Abner Doble's Model E

Occasionally, an inventor becomes so desperate they resort to unscrupulous methods to keep their invention alive. Abner Doble is such a case. Yet hearing his story, you can't help but feel sympathy for the fellow.

When Henry Ford introduced his Model T in 1908, its simple but reliable engine helped make it the best-selling car in the United States. It also spelled the end of steam-powered automobiles. By 1924, the Stanley Motor Carriage Company and Locomobile, two of the leading manufacturers of steam-powered cars, had closed their doors, while the White Motor Company had long since switched to selling cars with an internal combustion engine. What then was Abner thinking when he introduced his steam-powered automobile that same year? The answer is simple. Abner Doble's steam-powered Model E not only surpassed cars of its day in range, speed and acceleration, it continues to outperform many cars today.

ooo

Abner Doble began his mechanical apprenticeship aged 8, working in his grandfather's San Francisco machine shop after school. In 1906, the same year a Stanley Steamer set a land speed record, the 16-year-old collaborated with his three younger brothers to build a steam-powered automobile in their parents' basement.

When Abner was a freshman at MIT, he visited the nearby Stanley Motor Carriage Company to sell them on his new, improved steam condenser. The company ignored Abner's suggestion, but he was so confident in his design he dropped out of college and, with the financial support of his parents, opened a machine shop. The result was the Model A, a steam-powered

Warren and Bill Doble with the steam-powered car they built from spare parts in their parent's basement with help from Abner while still in high school. (UC, Berkeley, Bancroft Library)

automobile that Abner built with his brother, John. The Model A was followed by the Model B, which Abner drove all the way to Detroit to raise money for the next iteration. Forming the General Engineering Company with $200,000 in investor capital, Abner and his brothers rethought every aspect of steam technology, resulting in the Doble-Detroit.

When the Doble-Detroit debuted at the National Automobile Show in 1917, the automotive press hailed it as 'strictly up to date'. Sales materials boasted the car could travel 1,200 miles on a 25-gallon tank of water. It also had only four controls: two pedals on the floor (one for braking, the other for reverse) and a steering wheel with a speed throttle at its centre. Orders flooded in.

Abner always blamed the company's failure on the First World War when steel was diverted to the war effort, but the truth is more complicated. The Doble-Detroit was an improvement over the Model B, but proved sluggish and unreliable. Abner could also be a publicity hog. John, the most mechanically gifted of the Doble boys, bristled at Abner taking credit and sued him for patent infringement. When Abner's father sided with John, the family split in half.

Abner quit the car business after that. But John's premature death at age 28 from lymphatic cancer led to a reconciliation among the brothers. That year they founded the Doble Steam Motors Corporation in San Francisco and began working on the Model E.

Certainly, the Model E's performance was impressive. For example, it could travel 1,500 miles on a 17-gallon tank of water.* In contrast, a Model T's range was 200 miles. Additionally, a Model T needed forty seconds to reach its top speed of 40mph compared to the Model E, which accelerated smoothly from 0 to 75mph in less than ten seconds. As a result, the Model E could outrun any Duesenberg on the road, and did so as quietly as an electric car today.

Doble's secret was a revolutionary flash boiler smaller and more powerful than anything previously invented. It

* The Model E also required 26 gallons of kerosene to heat the water in its boiler.

not only built 750lb of steam pressure, it did so quickly. Most steam-powered cars required thirty minutes to build a head of steam before they could be driven. In contrast, the Model E had an electric starter that, with a turn of the key, heated the water in ninety seconds. Additionally, the kerosene used to boil the water burned at such a high temperature the car's only emission was water vapour. Today, the Model E would have no trouble passing California's smog test, the strictest in the nation.

Abner at the wheel of his steam-powered car.

Doble's Model E was remarkable for having only twenty-two moving parts. And since there was no transmission, there was no clutch or gear shift. Meanwhile, its four-cylinder engine came with a 100,000-mile warranty. No wonder the *Los Angeles Evening Express* called the Model E 'a motorcar that has shattered all conception of automobile performance'.

Abner's personal
Model E.

A Doble Model E
Deluxe Phaeton A.

Priced at $19,000 ($250,000 today), the Model E was squarely targeted at the luxury market.* Hollywood stars such as Buster Keaton, Wallace Beery and Norma Talmadge all owned one. Nineteen-year-old Howard Hughes was clocked doing 133mph while driving his Model E in the Texas desert. The Maharaja of Bharatpur even had special taps installed in his Model E to dispense beer and ice water while he was tiger hunting.

But the skills that make for an inspired inventor rarely include the ability to scale a venture. Doble Motors built each Model E by hand, one at a time. Abner also insisted on tinkering with every car before it left the shop, meaning no two Model Es were alike. Abner's insistence on using only the finest, most expensive components (the steering wheel was made of ebony, the headlights had components from Zeiss and Bausch & Lomb) ensured every Model E lost money. In other words: death by incrementalism, making Abner Doble an excellent example of the Fourth Rule of Failure: successful inventors often suck at mass production.

There's an important difference between inventors and commercialisers. Thomas Edison was a commercialiser; Nikola Tesla was not. Henry Ford was a commercialiser; Abner Doble, despite his best efforts, was not. In Abner's case, perfect was the enemy of good enough – an important distinction when it comes to controlling production costs. Sadly, this was just one of his shortcomings.

Doble Motors had no trouble raising a million dollars via a stock offering in 1923. But Abner's perfectionism drove manufacturing costs so high, no economies of scale could be realised. Desperate for money to keep his venture afloat, Abner sold $400,000 in additional stock he had no legal right to issue.

It wasn't long before Abner was indicted on five counts of security violations. Standing trial in San Francisco, he must have made a good impression upon the jurors because they asked the judge to show mercy. The judge disagreed, sentencing Abner to between one and five years in San Quentin. The California Supreme Court reversed the decision on a

* For perspective, a Model T sold for $260.

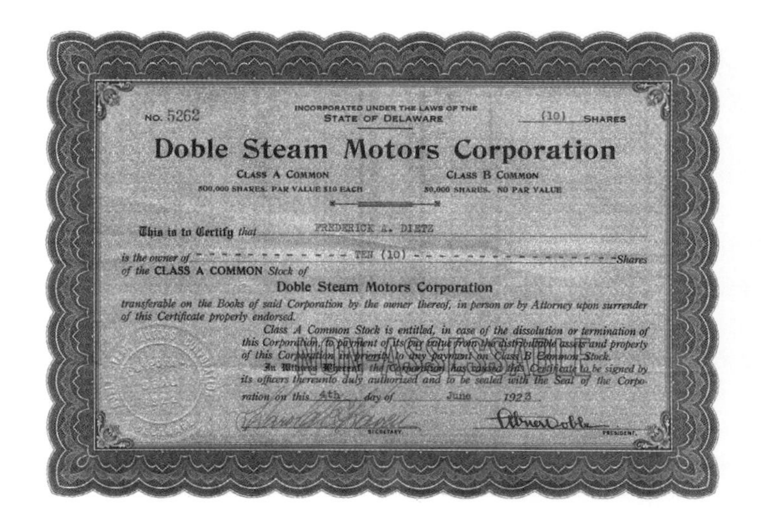

No. 5262

INCORPORATED UNDER THE LAWS OF THE
STATE OF DELAWARE

(10) SHARES

Doble Steam Motors Corporation

CLASS A COMMON
500,000 SHARES. PAR VALUE $10 EACH

CLASS B COMMON
50,000 SHARES. NO PAR VALUE

This is to Certify that FREDERICK A. DIETZ

is the owner of TEN (10) Shares
of the CLASS A COMMON Stock of

Doble Steam Motors Corporation

transferable on the Books of said Corporation by the owner thereof, in person or by Attorney upon surrender of this Certificate properly endorsed.

Class A Common Stock is entitled, in case of the dissolution or termination of this Corporation, to payment of its par value from the distributable assets and property of this Corporation in priority to any payment on Class B Common Stock.

NON-ASSESSABLE

In Witness Whereof, the Corporation has caused this Certificate to be signed by its officers thereunto duly authorized and to be sealed with the Seal of the Corporation on this 4th day of June 1923.

SECRETARY.

Abner Doble
PRESIDENT.

A Doble stock certificate. (Bancroft Library, UC Berkeley, Abner Doble Papers, 1885–1963)

technicality, but the damage was done. Doble Motors, already teetering on the brink of bankruptcy when the Depression struck, managed to stay afloat until 1931. That's when Abner was forced to liquidate its assets. In total the company had built and sold only forty-one cars, ten fewer than well-known automobile entrepreneur Preston Tucker.

Abner devoted the rest of his life to evangelising steam. Travelling to New Zealand, Germany and Britain, he consulted for local ventures that built steam-powered buses and trucks. In 1951, the chainsaw magnate Robert Paxton McCulloch hired Abner to develop a steam-powered solution for his experimental car, the Paxton Phoenix. McCulloch got as far as transplanting the engine from Doble's personal Model E into a 1953 Ford before the project was scrapped.

Abner died of a heart attack in 1961 alone and unheralded. He may never have fulfilled his steam dream, but he'd have been happy knowing that when a Model E comes up for sale they go for $1 million. Not such a failure after all.

35 The Besler Brothers and Their Steam-Powered Aeroplane

Abner's story doesn't end there, however. His steam technology was both innovative and sound. Surely, there was a money-making application. What's surprising is that someone thought it was in an aeroplane.

The Besler brothers, William and George, were hired in 1929 by Abner to work at Doble Motors. The brothers had a lot in common with their boss. The Besler family owned a financial interest in the Davenport Locomotive and Manufacturing Corporation, which built steam-powered locomotives. Additionally, their wealthy father was the chairman of the New Jersey Central Railway, which also had plenty of steam-powered locomotives.

George Besler, the oldest by two years, majored in geology at Princeton, while William had an engineering degree from the same institution. After joining Doble Motors, the brothers worked alongside Warren and Bill Doble on reducing the weight of a Doble flash boiler, so it could be used in an aeroplane.

Abner had believed since 1917 that steam could be used to power an aeroplane. Now, he was determined to prove it. When the financial situation worsened at Doble Motors, the Besler boys contributed their own funds to keep the company afloat. But an ugly dispute broke out between Abner and the Besler boys, so Abner ousted them from Doble Motors. The Besler brothers ended up suing Abner, claiming he was bankrupting the business. When the Beslers prevailed in court, Doble Motors was forced into receivership, allowing William and George to wrest control from Abner.

A new company was formed called Besler Systems with George as business manager and William as chief pilot and engineer. Despite their dispute, the brothers continued working with Abner on building a steam-powered aeroplane. Finally, after three years of experimentation, the Beslers arrived at Oakland Municipal Airport on 17 April 1933 in their steam-powered Buick to unveil the world's first steam-powered aeroplane.

A Besler steam aeroplane.

Photographers and newsmen mobbed the two brothers as they showed off their plane. While newsreel cameras rolled, Bill appeared jovial and easy-going in his leather flying helmet, although George looked far more serious in his three-piece suit and spectacles.

With his trench coat flapping in the wind, William climbed onto the plane's lower wing before settling in the cockpit. Then, when he turned a switch to ignite the boiler, the plane sprang to life.

It took only five minutes for the converted Travel Air to generate enough steam to be ready for take-off. Nor was there any need to hand start the propeller; it turned on its own. As reporters watched the two-seater plane speed down the runway, a white ribbon of water vapour trailed in its wake.

At least one observer expressed concern that the aeroplane wouldn't generate enough power to get airborne. But all doubts ceased once the plane, with Bill Besler at the controls, took to the sky and banked over San Francisco Bay.

People were accustomed to the ear-splitting noise of an aeroplane engine. But the Besler aeroplane was so quiet reporters on the ground could hear Bill Besler ask them, 'Well? How does it look?' as he flew by.

The Besler aeroplane engine was truly remarkable. The reciprocating, two-cylinder, V-type engine weighed only 480lb (300 of it boiler). Once steam pressure reached 1,200psi, the engine delivered up to 150hp, turning the propeller at a healthy 1,625rpm.

Incredibly, the aeroplane's 10-gallon water tank gave it a range of 400 miles. And since the engine burned only 40 cents of fuel oil for every 100 miles of flight, it was not only economical but one of the 'greenest' aeroplanes ever to fly.

Bill took off and landed three times that day, remaining aloft for a total of fifteen minutes.

'We have proved the practicability of the steam power plant in the air,' he told reporters afterwards. 'A steam propelled plane can be operated successfully and at much lower cost than one driven by internal combustion.'

The Besler flight made headlines across the nation. Newsreels showing the Besler brothers played in cinemas around the country, while *Popular Science* noted, 'Experts are watching (their) progress … with keen interest.'

But neither Besler boy had any interest in mass producing the world's first steam-powered aeroplane. They'd only built and flown one as a publicity trick to promote their steam technology. The brothers were smart enough to recognise it was too late to wean the aeroplane industry off the internal combustion engine. Instead, they wanted to target an industry more reliant on steam.

The Beslers went on to make a fortune selling their steam technology to the railways. Three years after their historic flight, they designed and built the Blue Goose passenger train for the New Haven Line.

Sadly, the Besler brothers' steam-powered aeroplane was destroyed in a storm. Their steam-powered aircraft engine was also lost to history after being sold to Japan in 1937. Still, they prove the point that the killer application of a revolutionary new invention isn't always what the inventors intended. This is why remaining flexible can make the difference between success and failure.

In summary, Moller, Cooley and Doble are good examples of the Fifth Rule of Failure: nothing goes as planned. But that doesn't mean they were unscrupulous, just blinded by belief in their invention.

Bill Besler built a replica of the original Besler steam-powered aeroplane engine and donated it to the Smithsonian. It is now part of the National Air and Space Museum's collection in Washington DC.

Chapter Seven

Getting There is Half the Fun

> I have not failed. I've just found 10,000 ways that won't work.
>
> Thomas A. Edison

Well before engineers tried shoehorning atomic energy into anything that would float, fly or drive, they tried slapping a propeller on it. Just like atomic power was all the rage, WETech inventors thought adding a propeller to any mode of transportation beside aeroplanes could solve a problem. More often than not it created new ones. This leads us to the Sixth Rule of Failure: just because something works well in one application doesn't mean it will work in another. Take cars for instance.

36 Propeller-Driven Cars

Propeller-driven cars didn't do any better on the road than they did in the desert. Still, inventors came up with many different examples of prop-driven cars with no intention of flying. One of the most concerted efforts took place between 1913 and 1926 when a French engineer named Marcel Leyat began experimenting with a propeller-driven car called the Hélica.

Leyat had experience building aeroplanes, whose base principles he thought suitable for a car. He began with a three-wheeled vehicle he called the Helicycle. Built in 1914 with two wheels in front, one in back and a front-mounted propeller that pulled the car forward, it proved highly unstable. After the Helicycle had a series of accidents, Leyat dumped the design in favour of a four-wheeled version, which he called the Hélica.

Leyat sold his first prop-driven car to the public in 1919. Nicknamed *L'avion sans ailes* (plane without wings), the Hélica was a simple machine. Instead of an engine turning its wheels,

The Leyat Hélica. (Courtesy Lane Motor Museum)

it turned a front-mounted propeller that pushed the car through the air. This meant the Hélica required no transmission, gearbox or clutch. Indeed, some prop-driven cars dispensed with brakes altogether, relying instead on putting the propeller in reverse to stop the vehicle.

A front view of the Leyat Hélica.
(Courtesy Lane Motor Museum)

The Hélica was slow to accelerate but once it got going it could be pretty fast. One was clocked at more than 100mph. Leyat's invention was also versatile. There was a sport model that looked like the open cockpit of a biplane with the driver sitting in front and the passenger behind him. There was also a saloon version with an enclosed cabin. One was even adapted for use on railway tracks in the French Congo.

Commercial production was small – fewer than thirty were built, but that's more than anyone else ever managed for a prop-driven car. However, there were several disadvantages. First, the Hélica steered via its rear wheels, with mixed results. Next, it was noisy as hell. It didn't help that passengers experienced a continuous torrent of wind in their face, not to mention the occasional diced bird. Although the Hélica's propeller had a guard around it, it provided only rudimentary protection, making it a menace to pedestrians. Despite this, Leyat can lay claim to the world's best-selling propeller-driven car, even if he sold only twenty-three of them.

○○○

Believe it or not, one of the earliest prop-driven cars was also meant to be amphibious. It was built by François Garbaccio, an Italian who immigrated to Switzerland. He repaired bicycles for a living before being bitten by the aviation bug. In 1911, he flew one of Switzerland's first aeroplanes. Built from bicycle parts, it didn't fly very high, far or for long.

It's unclear whether Garbaccio built his prop-driven car before or after he built his aeroplane. What is clear is that it had a huge, front-mounted propeller 6½ft long. It also came with flotation tanks mounted on the vehicle's side and rear. Garbaccio made extravagant claims for his invention, saying it would attain 60mph on land and 30mph on water. There is no record of it having attained either. In fact, only one photograph exists showing Garbaccio with his invention. Given prop-mounted cars do poorly on an incline, it's amazing he built it in the land of the Alps.

○○○

Leyat continued experimenting with various models of his propeller-driven car. In 1932 he built the Helicron, which looks like a giant, wooden wine cask with a cutting blade mounted in front.

The Helicron's controls were as simple as the Hélica's. There was a steering wheel, throttle and emergency brake. It also had a vertical stabiliser where the boot would be, which seems

Leyat Helicron. (Courtesy Lane Motor Museum)

A rear view of the Leyat Helicron. (Courtesy Lane Motor Museum)

unnecessary but added to the car's plane-like feeling. Because prop-driven cars are slow to build traction, the Helicron required the occasional push to get started. The most amazing thing about the Helicron is that, after being lost for sixty years, the sole existing model was rediscovered in a barn in France. It had to be completely rebuilt, but is now in working order and occasionally taken on demonstration drives. Even more amazingly, French safety inspectors deemed it road safe.

OOO

Jim Bede was another high-flying inventor who couldn't keep out of trouble. He started out designing and selling 'kit' aeroplanes. His most famous was the BD-5J;* a single-seat, micro jet made famous by its appearance in a James Bond movie.

Like a lot of inventors, Bede ran into trouble when it came to financing. Fourteen crashes of the BD-5 resulting in nine fatalities didn't help. When Bede Aircraft went belly up in 1979, the Federal Trade Commission discovered he had been using $9 million in customer deposits for other inventions rather than funding the aircraft kits he'd sold them. Bede signed a consent decree forbidding him from accepting deposits on aircraft for ten years. That's why Bede, like Dr Moller, is a good example of the Seventh Rule of Failure: when an inventor runs short of money they become desperate. They also never give up. Two years later, Bede went into the business of selling propeller-driven cars to the public.

The Bede Car used an 80hp motorcycle engine to drive a ducted fan. Putting a pusher propeller in the rear of the car eliminated the problem of mowing people down unless the car was backing up or got rear ended. With a top speed of 100mph, it was said to get 120 miles per gallon on the motorway, largely because it weighed only one quarter of a typical SUV.

Bede intended to sell his car in kit form for $8,000 beginning in 1982. Mass production was to follow four years later. He even took out a series of ads claiming the Bede Car was so fuel efficient that if everyone drove one, 'the United States could become an oil exporting nation'.

But the prototype didn't work as advertised. The car lacked power at low speeds, and had difficulty climbing even small hills – characteristic drawbacks of every prop-driven car. The

* Also called the Acrostar jet.

fuel rating also proved optimistic. Only one Bede Car appears to have been built, although three prototypes were claimed. What happened to them remains a mystery. Jim Bede himself died of an aneurysm in 2015 aged 82, selling aeroplane kits right up to the end.

There are two Hélicas still in existence: one is at the Conservatoire National des Arts et Métiers, in Paris, France; the other is at the Lane Motor Museum in Nashville, Tennessee. The Lane Museum also has a Helicron. A second Helicron can be found at the Automobile and Fashion Museum in Málaga, Spain.

37 Propeller-Driven Snowmobiles

One place besides the air that propellers do well is the snow. That's why Russians have been building snowmobiles with rear-mounted propellers for over a hundred years. Called aerosani, which translates as air sled, these ingenious vehicles are used in Russia's remote, snow-covered regions to carry passengers; deliver mail, medical assistance or emergency aid; and patrol the border.** They may have limited application, but they work so well they eventually found their way into combat operations.

The world's first aerosani may have been invented by accident in 1903 when an aviation engine was mounted on a sled for testing. One of the engineers, Sergey Nezhdanovsky, liked the idea so much he patented it. In 1910, Igor Sikorsky, who would one day become the father of the helicopter, built two aerosani equipped with a rear-mounted propeller for dashing through

** Soviet Premier Leonid Brezhnev used one as his personal hunting vehicle.

Kiev's snow-laden streets. That same year, the Grand Duke of Russia hired Franco-Romanian inventor Henri Coandă and the Paris-based powerboat building firm of Despujols to design a sleigh powered by a ducted fan. Coandă had a habit of exaggerating his accomplishments, but he did equip a custom-built wooden fuselage on runners with a front-mounted ducted fan. In one of the few existing photographs, a Russian Orthodox priest blesses the odd-looking vehicle, which has since disappeared.

The Grand Duke of Russia's aero sled.

One of the most successful series of aerosani was developed in the 1920s and '30s by Andrei Tupolev, a major contributor to Soviet aviation. The Ant I to V were used to carry passengers and mail on a regularly scheduled route between Russian cities. Initially fitted with automobile engines, they later used aircraft engines. Tupolev even made an amphibious version. The ultimate in hybrids, it travelled over snow, ice and water.

One reason aerosani are so useful is that they don't need roads, so can go places a car can't reach. This was important in the Soviet Union, where there were few roads in its remote, snow-covered regions. They're also cheaper to operate than a plane, and don't require expensive infrastructure like runways. But aero sleighs have some of the same problems as prop-driven cars. They're not only noisy, they have difficulty steering and climbing hills. As a result, they keep to frozen rivers and lakes, which are level and devoid of hairpin turns

Aerosan were also used by the Russian military during the First World War. In 1939, the Red Army used an updated version equipped with armour plating and a rear-facing machine gun for guerrilla raids against Finland. Not surprisingly, the Finns had their own version. During the Second World War, the Red Army equipped special units with two-man aero sleds to harass German forces. Predictably, the Germans responded in kind.

Soviet aerosani have continued to evolve over time. At least 100 Sever-2 (*Sever* is Russian for north) were built to deliver mail to remote Soviet outposts between 1958 and 1961. Frequently operating in temperatures around -50°F, they used the 'Pobeda' passenger car body, making them look like a 1940s Buick with a huge aircraft engine on their boot.

The Sever-2 was replaced in the 1960s by the more powerful Ka-30. The Ka-30 is remarkable for looking like an old Volkswagen bus with skis instead of wheels and a huge engine nacelle with a propeller at the back. Able to travel at speeds up to 60mph, there was a customised version for the KGB that transported up to eight people through Siberia's frozen wasteland.

Profile of a Soviet 02SS aerosani tank, also called the TsKB-50 armoured aerosani. Note the rear propeller and front-mounted cannon.

Two views of a Soviet 02SS aerosani tank's rear-mounted propeller.

To be fair, propeller-driven snowmobiles haven't failed in the marketplace so much as succeeded within a very narrow niche. That's why aero sleds continue to be built today in places like Canada and Norway. Russia remains the leader, however, with roughly 90 per cent of the market.

Soviet KA-30.

At least 100 Sever-2s were built to deliver mail to remote Soviet outposts between 1958 and 1961.

38 Propeller-Driven Trains

As if propeller-driven snowmobiles aren't exotic enough, consider the propeller-driven train.

One of the first built and tested was in Germany during the First World War when an engineer named Dr Otto T. Steinitz collaborated with Carl Geissen to mount an aircraft engine on a railway car and run it around a test track at the German Aviation Research Institute in Berlin. The test vehicle reached a top speed of 97mph – a considerable achievement given the best a steam locomotive could do at the time was 65mph.

Dr Steinitz was encouraged enough to follow up this success by outfitting a railway freight wagon with two aircraft engines. Known as the Dringos Wagen, the two-axle railcar had an aircraft engine mounted at either end, each with a two-bladed propeller.

The Dringos Wagen made a 25-mile test run from Grunewald to Beelitz and back in 1919. Roughly forty VIPs including railway crew boarded the small, box-like shed situated between the two props. Given that the railway car was outfitted with twin, six-cylinder aircraft engines each generating 160hp, it was hoped the Dringos Wagen would reach a top speed of 80mph.

The Dringos Wagen was a railcar with aircraft engines at both ends.

German engineer, inventor and patent attorney Otto Steinitz standing on the Dringos railcar he designed (third from left, wearing a hat and fur-collared coat).

But, as is often the case with WETech inventions, reality fell short. The Dringos Wagen was not only slow to accelerate – a common problem for anything prop-driven – it failed to top 40mph given concern about the car's structural integrity and primitive brakes.

There was hope the project could be taken forward, but that same year the Treaty of Versailles banned Germany from manufacturing aircraft engines for the military. The ensuing shortage combined with rampant inflation and uncertain financing nipped the project in the bud.

Despite the less than conducive conditions, Geissen continued to experiment with propeller-driven trains. As for Dr Steinitz, he was Jewish and was forced to flee Germany in 1939 along with his wife and children. Immigrating to New York City, he spent the rest of his life there until he died in 1964 aged 77.

Two years after the Dringos Wagen made its test run, an obscure Russian railway employee named Valerian Abakovsky convinced his bosses at the Tambov Railway Workshop to let him build his own version. Called the Aerowagon, it was a single railcar with an aircraft engine mounted on its front along with a wooden, two-bladed 10ft-long propeller.

As surprising as it may seem to mount an aircraft propeller on a train, multiple inventors often come up with the same invention at the same time. As Malcolm Gladwell noted in his *New Yorker* article 'In The Air: Who Says Big Ideas Are Rare', the phenomenon of simultaneous discovery 'turns out to be extremely common'.

Abakovsky had strong ideas about the design of Russia's first high-speed train. For example, he shaped its nose like a wedge to reduce wind resistance. He also envisioned it as an exclusive means of speedy transport for high-level Soviet officials.

The Aerowagon underwent extensive testing in 1921, travelling nearly 2,000 miles at speeds up to 87mph. When it was finally deemed ready, its first official act was to take delegates from the Communist Internationale meeting in Moscow on a VIP tour. Heading up the delegation was a close friend of Joseph Stalin's. Other guests included foreign delegates from Australia and Germany as well as Abakovsky, the inventor.

The tour started well enough. After picking up the delegation in Moscow, the Aerowagon safely delivered them at moderate speed to a coal collier outside Tula, then to the Tula weapons factory. But when the VIP tour was finished, the delegation wanted to return to Moscow in a hurry. That's when the Aerowagon was put to the test.

After being cranked up to its maximum speed, Abakovsky's invention promptly derailed, disintegrating in the process. Of the twenty-two people on board, six were killed instantly, including Abakovsky who was only 25 years old. A seventh died later of his injuries. It goes without saying that further development of the Aerowagon died with them.

Russian Aerowagon engineer Valerian Abakovsky.

Abakovsky's Aerowagon.

The same year that Abakovsky met his ignominious end, a wealthy Scottish inventor named George Bennie patented the Bennie Railplane. A streamlined railcar with a propeller on either end, the Railplane was granted the first patent in the UK to use 'Aerial Tracks for Guiding Aircraft'. Note the patent says 'aircraft' not trains.

Bennie's Railplane was certainly revolutionary, but it was more monorail than plane. Suspended from an overhead rail with a stabilising track beneath to keep it from swaying, the canister-shaped coach with a propeller at either end was memorable looking. Dual electric motors drove its twin, two-bladed propellers at speeds promised to reach 120mph. Braking was achieved by slowing the wheels attached to the overhead rail while the propellers were put into reverse.

Bennie began his Railplane venture in 1921, proposing a mile-long connection between two London neighbourhoods. The necessary permissions were granted, but the project ground to a halt due to local opposition and a lack of funds. Despite these setbacks, Bennie was determined to build a full-size prototype. Finally in 1929, he got approval to build a test track just outside his home city of Glasgow.

The Bennie Railplane was intended to straddle existing railway lines on a huge, lattice-like structure made of metal that towered above the landscape. While the Railplane zipped overhead, providing a high-speed rail service to passengers who could afford it, slower-moving freight would travel by train on the conventional tracks below.

The Railplane's coming out ceremony took place in July 1930. One hundred and forty VIPs and invited guests including Bennie's mother and sister arrived to ride in the world's first propeller-driven monorail. Men in suits wearing stiff collars and smoking cigars climbed a flight of stairs to a platform high above the Scottish countryside, where they were greeted by Bennie wearing a raincoat and bowler hat. Nodding affably, he spoke a few words to explain his invention before they boarded the train. A steam locomotive parked underneath the platform made it clear the future of transportation had arrived.

A proposed advert for the Bennie Railplane, c.1930. (University of Glasgow Archives & Special Collections, DC85)

Bennie's demonstration coach, which carried up to twenty-four passengers, was outfitted with plush carpeting and curtains along with comfortable seats and small tables. Six large, rectangular windows on either side of the coach (plus two smaller portholes) provided excellent views. The doors at either end boasted an oval window with panes of bevelled glass. If there was a drawback, it was the cramped interior. Many support columns made the carriage feel narrow, especially when filled with passengers. Bennie probably thought it cosy.

The test track may have been high in the air but it was less than 500ft long, making for a short ride. Still, it was an impressive demonstration. As one rider told the *Glasgow Herald*, the Railplane 'operated with perfect smoothness and passengers only knew the car was moving by gazing out of the window at the passing landscape'.

After the festivities, Bennie turned his Railplane into a local attraction. Anyone with a shilling could afford to take a ride. But despite worldwide media attention, no investors stepped forward

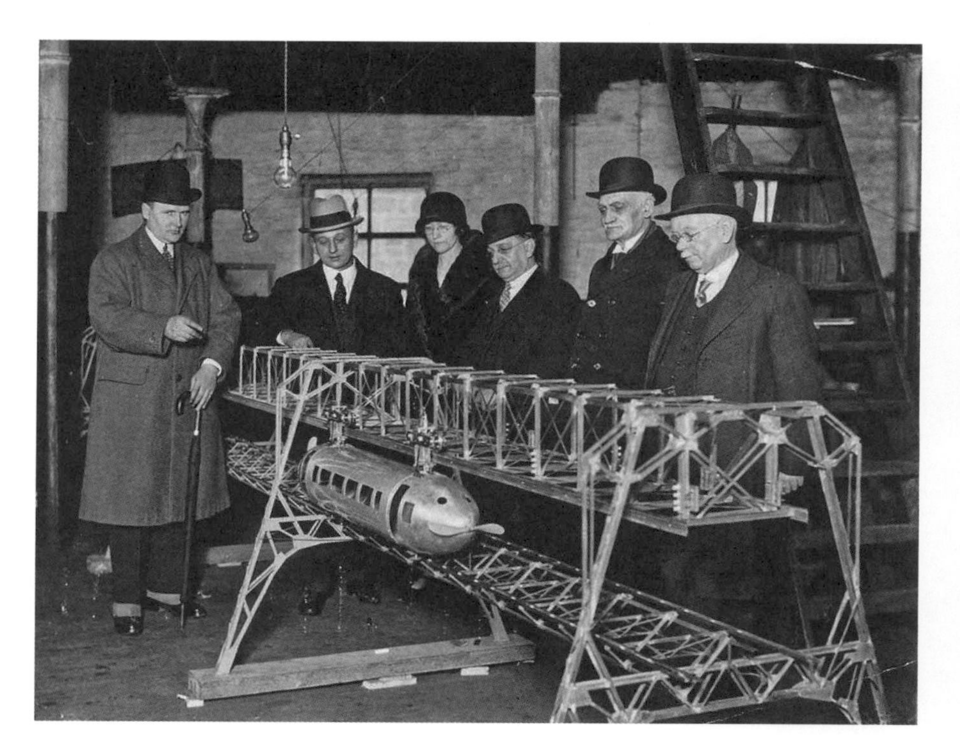

George Bennie (left) demonstrating his Railplane model to interested parties, *c.*1930. (University of Glasgow Archives & Special Collections, DC85)

to commercialise his invention and the attraction closed after only two months.

The Railplane's economics combined with the Great Depression sank the project. The Railplane required a lot of expensive infrastructure built from scratch. And although existing railways could provide the right of way, they saw Bennie more as a competitor than a profit-making partner. Without their co-operation, the cost of land acquisition made Bennie's project too costly to pursue.

Since Bennie financed much of the project himself, he ran out of money. In 1936, he resigned from the company he'd founded (some reports suggest he was ousted), declaring personal bankruptcy the following year.

Sadly, the Bennie Railplane did not survive its inventor. Its Scottish test track was dismantled in 1941 and sold for scrap to aid the war effort, while the Railplane lay rusting in a field. Although remnants of the track can be found today, the actual Railplane (except for a scale model) has been lost to time.

Bennie continued working on his invention, albeit alone, proving just how determined he was to realise his dream. In 1946, he founded George Bennie Railspeed Ltd and got permission to build a 4-mile test track outside Glasgow until the city council changed its mind. In 1950, Bennie visited Syria and Iraq, where he proposed linking Baghdad and Damascus by aerial train. Despite his determination, nothing came of these efforts. Bennie, who never married, ran a herbalist shop in Scotland after that. He died in obscurity in a nursing home in 1957. He was only 65 years old.

ooo

While Bennie was demonstrating his Railplane in Scotland, a German airship engineer named Franz Kruckenberg was working on an experimental railcar with a rear-mounted propeller for the company he'd founded. Called the Schienenzeppelin, or 'rail zeppelin', Kruckenberg's train looked like a Zeppelin airship right down to the erector set-like skeleton beneath its aluminium skin.

Built near Hanover, Germany, in 1930, the Schienenzeppelin was soon ready for testing. Driven by a single, 9ft-long, downwards-tilting propeller,* the Schienenzeppelin was amazing to behold. Built on a steel chassis, its truss frame was perforated by holes to lessen its weight. It was also streamlined to reduce wind resistance. Its front was round-nosed, while its rear ended in a Y-shape from which a whirling propeller extended.

The 52ft-long Schienenzeppelin could hold between twenty-four and forty-four people depending on its configuration. BMW engines drove the propeller; the same engine type that would power the Luftwaffe's light bombers. The coach interior, done in Bauhaus style, had wood panelling and included a bathroom for passengers. Windows ran the length of the carriage, offering plenty of light, but none opened due to the Schienenzeppelin's anticipated speed.

---

* Initially, it was a four-bladed propeller; later, a two-bladed one, both made of ash.

Schienenzeppelin. (Franz Jansen)

The first propeller-powered test took place on 25 September 1930. Nine months later it hit 143mph on a test run between Karstadt and Dergenthin – a new speed record for passenger rail travel that stood until 1954.* Given the fastest streamlined locomotive topped out at 120mph, the Schienenzeppelin was

* The Schienenzeppelin remains the fastest propeller-driven train.

a *wunderkind*. And since it only weighed 20 tons, it was more fuel efficient.

Tests revealed braking was an issue. In one case, it took more than a mile for the Schienenzeppelin to stop. It was decided that a fully reversible propeller would improve braking, but the problem was never resolved.

That didn't matter when it came to the public. The Schienenzeppelin was a crowd-pleaser. So much so, the police had to hold back spectators whenever it appeared. No German company would insure the propeller-driven train, however. Special arrangements had to be made with Lloyd's of London for coverage.

Despite its remarkable speed, the Schienenzeppelin never entered passenger service. One drawback: its propeller-driven design didn't allow for additional units to be coupled.** Additionally, it seemed imprudent to have a spinning propeller next to passengers as they waited on the station platform.

The prototype was sold to the German State Railway Company in 1934. While Kruckenberg worked on more conventional designs, the Schienenzeppelin languished in storage. Five years later it was disassembled and its parts used by the German military as it geared up for war.

39 Hover Trains

Not every WETech inventor works alone, underfunded and in obscurity. Take Jean Bertin.

Bertin had a gold-plated upbringing with sterling academic credentials. A member of the French elite, he graduated from the École Polytechnique before serving as an engineer in the French Air Force. From 1943 to 1955, Bertin worked at the French National Society for the Development of Aircraft Engines

** Bennie solved this problem by making the Railplane's rear-mounted propeller removable, so more cars could be added.

(SNECMA), where he rose to Deputy Technical Director. In 1956, he founded Bertin & Cie, which spent more than ten years perfecting the hover train.

If Bennie's Railplane was more train than plane, then Bertin's invention, the Aérotrain, was more plane than train. Convinced that aviation technology could improve rail transit, Bertin's goal was to build a hover train with a jet engine that glided along an elevated track on a cushion of air. This was a serious effort led by a formidable individual with deep reserves of common sense. Millions of francs were spent on the effort, 600 staff employed and impressive milestones achieved.

Bertin's Aérotrain capitalised on the ground effect principle in which friction is greatly reduced because the object floats on air, eliminating the need for wheels. Bertin intended his jet-powered passenger train to travel at high speed. He even employed an aerodynamic design to reduce wind resistance. Unusually inventive, he was used to getting results.

Four Aérotrain prototypes were built between 1965 and 1977: the 01, 02, S44 and I80. The 01 was a half-scale prototype, but large enough to accommodate two crewmen and four passengers. Originally equipped with a three-bladed, reversible-pitch propeller, it was later upgraded with a jet aircraft engine.

Bertin tested the 01 in 1966 on an elevated track 4 miles long in Gometz-le-Châtel. His first Aérotrain prototype reached a top speed of 214mph, proving the concept worked. Testing soon progressed to a second prototype, the 02, which looked notably different from its predecessor. The 02 resembled a rocket sled more than a passenger train. With three air intakes in front, an aircraft canopy over a two-seater cabin, and a pylon in the rear sporting a three-bladed propeller, the 02 was straight out of science fiction. A few adjustments, including a more powerful jet engine and a different hover skirt, improved the air cushion effect, helping the 02 reach a top speed of 262mph.

Testing continued with a third prototype, the boxy-looking S44. Unlike the 01 and 02, which had jet engines, the S44 was equipped with a linear electric motor rather than a propeller to drive it forward. Given the difference in technology, a second test track had to be built alongside the first.

The S44's electric motor was far quieter than the jet engines on the 01 and 02 – a plus when operating a suburban line. Additionally, it had little trouble accelerating and decelerating – something that was difficult for the propeller-driven Aérotrains to accomplish. But the S44 couldn't go nearly as fast as Bertin's previous prototypes, so a fourth Aérotrain was built for high-speed passenger service.

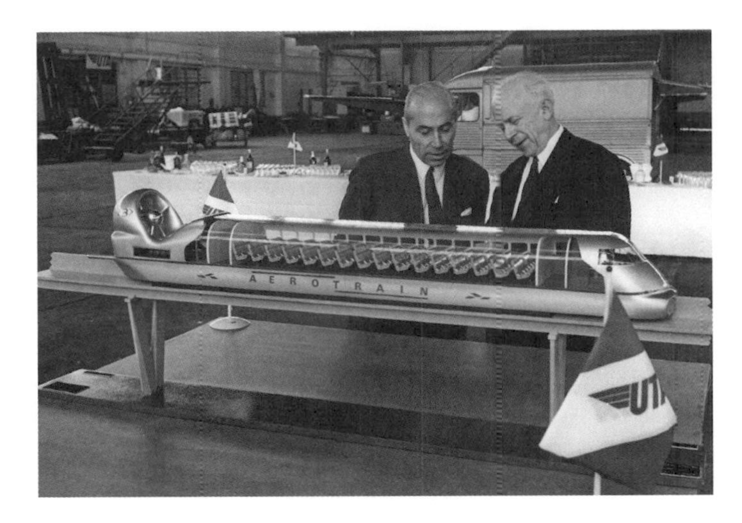

Jean Bertin (left) showing a model of the Aérotrain, July 1969. (Keystone France/Gamma-Rapho via Getty Images)

Yet another elevated test track, this one 11 miles long, was constructed in Loiret, France, to accommodate the Aérotrain I80-250. Striking to look at, it had a long, sleek, silver-coloured body made of aluminium, with a red stripe running the length of the car. The I80-250 was a full-sized passenger coach intended for intercity service. Designed to carry eighty passengers, its twin turbine engines turned a seven-bladed propeller that drove the Aérotrain forward, while twelve compressors created a cushion of air to float on.

The latest version of the Aérotrain reached a peak speed of 186mph, but Bertin was convinced he could do better. Upgrading the prototype, he renamed it the I80 HV for High Velocity and began testing in 1973.

Things seemed to be going well for Bertin. The Aérotrain I80 HV set a world speed record for its class in 1974, reaching a peak of 267mph. The French post office even released a stamp honouring the project. But the Aérotrain wasn't without challenges. For example, it was cheaper to upgrade existing railway tracks to handle a high-speed train than build a new elevated infrastructure for the Aérotrain. Additionally, Bertin's prototypes used gas turbines with giant propellers. This created a tremendous racket both inside and outside the train. They also consumed a lot of fuel – an expensive proposition during the 1970s fuel shortage.

The Aérotrain 01 Prototype, as seen in Paris, 2013. (Siren-Com/ CC 3.0)

Nevertheless, Bertin felt the technology was mature enough for implementation. In June 1974, the culmination of everything he'd been working on came to fruition when he signed a contract with the French government to build an Aérotrain transit line on the outskirts of Paris.

The railway industry is a powerful force in France. It did not look kindly upon a paradigm-shifting transportation system that not only threatened its profits, but was built by the aircraft industry. Less than one month after the deal was signed, the contract was cancelled and support shifted to the TGV, a high-speed train that ran on conventional tracks. This brings us to the

The Aerotrain S44 Prototype.

The Aerotrain I80 HV in 1969. (Yves Le Roux/ Gamma-Rapho via Getty Images)

Eighth Rule of Failure: entrenched industries will fight any technology that threatens their profits.

The setback was devastating for Bertin, who'd spent more than a decade promoting the Aérotrain. Bitterly disappointed and riddled with cancer, he stepped down from his company in 1975, dying shortly thereafter. He was only 58 years old.

The Aérotrain I80 HV made its last trip in 1977. After that it was mothballed in its hangar. But Bertin's achievements did not go unrecognised. He was made a Knight of the Legion of Honour, France's highest order of merit for a civilian. Additionally, Bertin Technologies lives on as a research and development firm doing contract work for the energy, space, defence and pharmaceuticals industries, but the Aérotrain is not among its projects.

Of the four Aérotrain prototypes built, only two survive. The S44 was destroyed by a fire in its hangar in July 1991. The next year, the I80 HV was also destroyed, this time by arson. The remaining Aérotrain prototypes 01 and 02 are in storage, but are occasionally displayed. For those caring to make a pilgrimage to France, Bertin's grave is located in Monblanc in Gers. The first test track for Aérotrains 01 and 02 can be found in Gometz-le-Châtel, although it's mostly in ruins. However, the S44's test track, which parallels the first, has been restored as a greenway for pedestrians and cyclists. The Aérotrain's third test track in Loiret has largely been demolished, but parts can still be found today.

40 Jet Trains

If a propeller can make a train go faster, just imagine what a jet engine could do! That's what the New York Central Railroad intended to find out during the summer of 1966.

The M-497, nicknamed the 'Black Beetle', was the world's first jet-powered train. Alfred Perlman, President of the New York Central, personally approved the project. After saving the railway from bankruptcy, Perlman was desperate to lure passengers back and hoped publicity around a high-speed passenger train might do the trick.

Revitalising train services in the United States was a tall order by 1966. The public had abandoned railways in favour of their personal cars, which were more convenient, or air transportation, which was quicker. Using a Budd Company railcar it had purchased in 1953, the New York Central sent it to its Ohio technical centre to be fitted with two surplus General Electric J-47 jet engines. Once used to power air force bombers, the jet engines were bought second-hand for the bargain price of $2,500 apiece.

The jet engines were converted to run on diesel fuel and mounted on the roof of the train at a slight downward-facing angle. A slant-nosed fairing, which looked like a welder's mask, was added to the front of the locomotive to reduce wind resistance. Additionally, more than fifty instruments were added to measure speed, stress and ride quality. A black paint scheme, designed by the wife of the project's engineer, gave the train its Black Beetle nickname.

The New York Central's jet-powered M-497 railcar, nicknamed the 'Black Beetle', rockets down the rails during its July 1966 test.

Testing took place on a July weekend along a 21-mile section of track. The track, running between Butler, Indiana, and Toledo, Ohio, had to be upgraded to accommodate the Black Beetle's anticipated speed – an expensive proposition if it was to be rolled out across the line. Railway crossing guards were lowered in advance and locked into place because the train moved so fast it would already have passed by the time the crossing guard came down. Railway ties were also placed across the end of the test track to derail the M-497 if it ran out of control.

The test was considered so important that the New York Central's President attended both days. Driving the train was the man who helped design and build it, Don Wetzel. Wetzel was the New York Central's Assistant Director of Technical Research. A former US Air Force pilot, he was also a licensed train engineer.

The Beetle ran two high-speed tests on Saturday, 23 and Sunday, 24 July 1966. You could see the heat waves coming off the jet engines as the train rocketed down the tracks, kicking up a cloud of dust in its wake. The Beetle's jet engines ran so hot their automatic shut-off feature had to be disabled. And the thing was so loud, trackside neighbours thought a jet fighter was making a low-level pass. Wetzel claims that he never once let go of the whistle cord during the entire 21-mile run. Whether it was out of fear or a concern for public safety, he never made clear.

It was hoped the Beetle would reach a top speed of 180mph. It hit almost 200mph instead – a world record for a light-rail vehicle travelling under its own power.

The Beetle proved that a high-speed rail service could travel on conventional train tracks, but its configuration was impractical. The Beetle's jet engines made the train so tall that it couldn't fit under railway bridges or through tunnels. Additionally, the jet engines had no reverse, so the Beetle had to be towed back to its starting point by the very type of diesel locomotive it intended to replace.

The New York Central never put the Black Beetle into production and removed its jet engines, which were then used to melt snow. The M-497 railcar returned to service, but two years later the New York Central merged with the Pennsylvania Road. Shortly thereafter, the combined company was acquired by Conrail, which eventually sold the M-497 to the Metropolitan Transportation Authority (MTA). The MTA cannibalised the M-497 for parts before selling it for scrap in 1984. Thus ended the world's first jet-powered train.

A few years after the Beetle, the Soviets tried something similar. In 1970, the Tver Carriage Works, a Soviet train factory with a

In 1971, the Soviet SVL reached a top speed of 160mph.

distinguished history, placed two AI-25 jet engines on the roof of an ER22 locomotive. Named the SVL for High Speed Laboratory Railcar, it was a rolling test bed for high-speed rail travel. The train reached 160mph in 1971, but the prototype was not commercially viable in part because of its high fuel consumption.

One thing that's clear from all this experimentation is that France, Germany, Japan and China have been far more adventurous when it comes to high speed rail than the United States. The question is, why?

Population density in Europe and Japan has something to do with it, as do smaller distances between major cities. A strategic commitment funded by generous government subsidies also contributes to their success. But when it comes to the United States, the only consistently profitable passenger rail line is the Boston–Washington corridor. Otherwise, the country is just too big, government subsidies too small, legacy infrastructure too old, or alternative means of transportation too readily available for a high-speed passenger-carrying railway to be profitable.

A futurist-style magazine cover showing the Soviet SVL.

Chapter Eight

Things That Don't Need Improving

> Success has many fathers but failure is an orphan.
>
> Common saying

WETech inventions often fail because they try improving upon something that's already close to perfect. How do you make a better mousetrap, bicycle, motorcycle or lawn mower? It's difficult, if not impossible. But try telling that to a WETech inventor who's convinced they have a better idea. What follows is a brief survey of inventions that failed because they tried improving upon something that didn't need it.

41 Sailboats

Sailboats have been getting people from one place to another thanks to wind power for thousands of years. But in 2008 world-class sailboat designer Ichiro Yokoyama teamed up with a professor at Japan's Tokai University School of Marine Science and Technology to build a wave-powered boat that took forever to get anywhere.

'Fossil fuel will run out one day, so I have studied wave propulsion as a promising way to save energy,' Tokai's Dr Yutaka Terao, a specialist in wave-powered boats, told *The New York Times*. Despite this assertion, it's not clear that wave power is an effective means of propulsion.

Configured like a catamaran, Dr Terao's boat was made of recycled aluminium in keeping with its green objective. But the *Suntory Mermaid II*, as it was known, had a wave-powered propulsion system located in its bow. The mechanism consisted of two horizontal fins side by side that moved up and down with the motion of the waves. Simulating a dolphin tail kick, the device was designed to pull the boat forward regardless of weather, wave height or wind direction. This eliminated the need for either wind power or an engine, though it had an outboard motor as well as a mast and sail in case of an emergency.

The *Suntory Mermaid II*, a wave-powered boat. (Kurita KAKU/Gamma-Rapho via Getty Images)

To prove the concept of a 'wave-devouring boat' worked in the real world, an ecologically minded sailor, Kenichi Horie (with two world sailing records to his name), captained the prototype from Honolulu to Japan, a distance of 4,300 miles. Horie completed the journey, but it took him three and a half months to do so, largely because the boat averaged less than 2mph. There are ocean currents that move faster. Since then, marine builders have shown little interest in commercialising wave-powered boats, anchoring them firmly in the category of WETech invention.

The *Suntory Mermaid II* is on display at the Tsuneishi Shipyard in Fukuyama City, Japan, where it was built.

42 Bicycles

The penny-farthing bicycle was all the rage when introduced in 1871, but it had obvious problems. First, it was difficult to balance on such a gigantic front wheel. Additionally, it was hard to pedal since the pedals were attached directly to the wheel rather than via a sprocket and chain as on today's bicycles. Given these drawbacks, there were many attempts to improve upon the design.

In 1882, Charles W. Oldreive invented a bicycle he called the New Iron Horse. Oldreive eliminated the penny-farthing's rear wheel while bracketing its big wheel on either side with smaller wheels to stabilise it. The rider sat inside the centre wheel rather than on top, using their hands to pedal. It's safe to say White Elephant Technology ran in the family since Oldreive's father is credited with inventing floating shoes that walked on water.

But it wasn't until 1885 that John Kemp Starley revolutionised the industry by inventing the safety bicycle. A chain-driven bike with two same-sized wheels, Starley's creation rapidly replaced the penny-farthing. The next big

Oldreive's New Iron Horse ridden by a woman on a country road in 1882. (Charles W. Oldreive/ Library of Congress)

innovation followed shortly thereafter. Called the tandem bicycle, it allowed one rider to sit in front with a second rider sitting behind.

The tandem proved successful in part because of the romantic possibilities of 'a bicycle made for two', as described in the song 'Daisy Bell'. But not all tandems were created equal. England's Star Cycle Company sold a tandem in 1896 that allowed both riders to control the steering. There was also a type of tandem called the 'sociable', or 'buddy bike', where riders sat side-by-side rather than front-to-back. It never caught on.

Given the safety bike's success on land, it wasn't long before people began building bicycles that travelled on water. By the end of the nineteenth century, inventions with names such as the water velocipede and hydrocycle could be found navigating New York City's Central Park Lake. In 1914, Parisians were organising hydrocycling races on a nearby lake with homebuilt craft of crazy inventiveness. These included a water tricycle with three giant, balloon-like tyres, and a two-wheeled hydrocycle powered by a propeller that spun not in the water, but in the air.

And then there was the Cyclomer. The Cyclomer made its first appearance at a Parisian trade fair in 1932. An amphibious bicycle that could be pedalled on land *as well as* water, the Cyclomer's wheels were giant hollow spheres with four smaller spheres used as outriggers to keep the vehicle afloat. Fins on the rear wheel acted like paddles driving it forward, or at least that was the idea.

A water velocipede, as seen in *Knight's American Mechanical Dictionary*. (NY: Hurd and Houghton, 1377; CC 4.0)

The Cyclomer, ridden by its inventor.

The inventor, E. Fabry, is said to have ridden the Cyclomer across a swimming pool with little trouble. There's even a drawing depicting him riding one on water while smoking a pipe. But the only known photograph of the Cyclomer shows the inventor sitting atop it on dry land in a stationary position – not exactly proof of concept. Other reports suggest its fins were too small to generate sufficient movement. Since the Cyclomer never caught on, it's fair to describe it as a flop.

A man operating a pedal-powered water tricycle, *c*.1900. (Bain Collection/ Library of Congress)

A variety of far more dependable hydrocycles are sold today, but some of the strangest bicycles ever created were the ones intended to fly.

An outbreak of flying bicycles took place in France in 1912, when Robert Peugeot offered a 10,000 franc prize for the first human-powered bicycle to fly a minimum distance of 33ft. Peugeot's family had started out in the bike-manufacturing business before turning to automobiles. The French found the notion of flying bicycles amusing and closely followed their development. They called them *aviette*, which means a human-powered aircraft.

The first Paris Pedal-Aeroplane Trial was held in June 1912. It attracted much attention, if achieving little success. Of the 198 entrants, only thirty-two showed up and none got airborne. The second trial was held in November that same year. *Cycling* magazine noted the flying bikes were somewhat improved, but still 'crude contraptions'. Among the fifteen competitors was a bike with a pedal-driven propeller, and a tricycle monoplane in which the aviator sat in a reclining position. There was even a bike with flapping wings. What there wasn't was a winner.

Peugeot amended the rules, but even then no one captured the Peugeot Prize for another nine years. Then in 1921, a champion cyclist named Gabriel Poulain pedalled a 37lb bicycle with

The Aviette Contest, Issy les Moulineaux, near Paris, France, 7 June 1912. (Jules Beau/ Bibliothèque nationale de France

This French contestant hoped to prove he'd invented the first flying bicycle. He hadn't. (Bain Collection/Library of Congress)

Gabriel Poulain on his 9 July 1921 record-setting run, which earned him the Peugeot Prize for human-powered flight.

a large upper wing and a smaller rear one at a speed of 28mph. When Poulain dipped the rear wing, his bike jumped into the air.

Poulain flew the prescribed distance four times, his longest flight being 38ft. Although he broke two spokes on one of his landings, his achievement earned him international fame and coverage on the front page of *The New York Times*.

○○○

France wasn't the only country interested in bikes that could fly. The United States also had aspirants. One advocate was a First World War pilot from Dallas, Texas, named Harry T. Nelson. In 1931, Nelson promised boys 'a wonderful chance to get into flying … Just send us 25 cents and get our easy plans for making a genuine Glide-O-Bike.'

One Nelson ad quoted a satisfied customer saying, 'Your Glide-O-Bike is the dandiest thing I ever owned. It sure gives you all the flying thrills you want.' The advertisements also suggested boys could recoup their investment by charging their friends 10 cents a ride.

The Glide-O-Bike conversion kit consisted of a set of wings and tail that could be attached to a bicycle. Nelson's ad promised,

'You can bank, ground loop, stall, and side-slip. Absolutely nothing like this for fun and thrills.' A list of aviation terms was included with the Glide-O-Bike's plans. What isn't clear is whether they explained that ground looping and stalling were two things you never wanted to do in an aeroplane.

An ad for the Glide-O-bike (Courtesy Nostalgic Reflections Museum)

Modern Mechanix reported that Nelson proved the safety of his invention by riding it in a gale, 'coming through the "flight" without accident or discomfort'. Eventually, Nelson's ads clarified that the only part of the bike to leave the ground was the front wheel, but that's hardly the impression they gave.

Human-powered bicycles that fly have come a long way since then. In 1979, Bryan Allen pedalled the Gossamer Albatross across the English Channel, winning the $200,000 Kremer Prize. The flight took under three hours, with Allen skimming just above the waves, averaging 8mph on the 22-mile flight.

Designed by Paul MacCready, a California-based aeronautical designer, the Albatross weighed only 60lb despite a 96ft wingspan. Although a fantastic accomplishment, the Albatross still qualifies as a WETech invention since there's no commercial market for the technology and only two were built.

The Gossamer Albatross II in flight. (NASA Photo)

The Gossamer Albatross I can be seen at the Smithsonian National Air and Space Museum in Washington DC. The Gossamer Albatross II is on display at the Museum of Flight in Seattle, Washington.

43 Submarines

Human-powered flight is one thing, but a human-powered submarine is quite another. Nevertheless, the Foundation for Underwater Research and Education (FURE) is dedicated to developing the technology. It's even sponsored the biennial International Submarine Race for more than twenty-five years.

Hosted by the Naval Surface Warfare Center, the International Submarine Race is a design competition to build and operate a human-powered underwater vehicle. Created in response to the shortfall of personnel in the marine engineering fields, the competition is designed to attract the world's best and brightest students.

The 14th International Submarine Race in the David Taylor Model Basin at the Naval Surface Warfare Center, in West Bethesda, Maryland, 29 June 2017. (Devin Pisner/ US Navy)

A human-powered submarine named *Umptysquatch 8*, courtesy of New Jersey's Sussex County Technical High School, one of twenty-four teams participating.

In order to compete, student teams build a human-powered submarine capable of navigating an underwater course. Since the one- or two-person subs are not water-tight, their occupants use scuba gear to breathe from an on-board air supply. The race ensures students gain real world experience building innovative propulsion and guidance systems, which sounds practical even if the outcome, a pedal-powered submarine, isn't.

44 Cargo Ships

In 1986, Captain D.C. 'Sandy' Anderson of Earth-Ship Ltd fitted a 3,500-ton grain carrier named the *Carib Alba* with an auxiliary wind-propulsion system called Comsail. That's right; a propeller-driven cargo carrier that relied on bunker fuel was outfitted with sails like a nineteenth-century schooner.

The *Carib Alba*. (Courtesy Captain D.C. Anderson/Earth Ship Limited)

'On a perfect day it saved an astonishing 35 per cent of fuel,' Anderson recalled.

But Anderson's experiment was short-lived. When oil prices collapsed, the *Carib Alba*'s owner took a blowtorch to its sailing masts, leaving them in a pile on a Honolulu pier. That was thirty-five years ago. Today, a handful of companies seek financing to develop their unique solution for wind-powered cargo ships, be it via sails, a Flettner rotor, wind turbines or a giant kite.

'There are a number of projects looking at the use of wind as a power source for shipping,' the technology editor at *Lloyd's List* told *The New York Times*. 'Whether these projects will prove successful ... remains a question.'

As it turns out, supplemental sail power is only suitable for cargo ships in the 3,000- to 10,000-ton range. However, there are twice as many container ships that can't avail themselves of wind power because they're too large and too fast. Nevertheless, the nineteenth century was the last time wind power was used to profitably transport goods. Since then it's become White Elephant Technology, which is ironic given how predominant it once was.

45 Lawn Mowers

After the push lawn mower, the most important innovation in lawn mowing was a mower powered by a two-stroke engine. After that came self-propelled mowers and the riding type. But in 1957, the Simplicity Manufacturing Company in Port Washington, Wisconsin, demonstrated what it called the 'Yard-Care Appliance of the Future'.

Simplicity needed a gimmick to promote its new line of Wonderboy lawn mowers. The answer was the Wonderboy X-100. The prototype was demonstrated on 14 October 1957, just ten days after Sputnik launched, which seems appropriate given what it looked like. For starters, the X-100 was steered by a flight stick similar to what you'd find in an aeroplane. Additionally, it came with headlights in case you wanted to mow the lawn at night. It also had a handheld radio for when you wanted to communicate with your spouse. The company said you could even run errands in it.

According to Simplicity's promotional materials, the X-100:

> has a five-foot diameter plastic sphere in which the rider sits on an air foam cushioned seat. It has its own electric generating system for ... air conditioning and even a cooling system to provide a chilled drink on a hot day. It can mow the lawn, weed it, feed it, seed it, spray for insects, plow snow and haul equipment. It can even be used as a golf cart.

No wonder the company called it 'the future of lawn mowing'.

The Wonderboy X-100 was so far ahead of its time it attracted media attention. It was even featured on the March 1958 cover of *Mechanix Illustrated*. Still, it's doubtful the X-100 was capable of everything the company claimed. One clue is that despite

The Wonderboy X-100. (Courtesy Briggs & Stratton)

being promoted as the company's flagship lawn mower, it wasn't for sale.

Simplicity was eventually bought by Briggs & Stratton in 2004, but the company still sells lawn mowers today. There are no futuristic lawn mowers in its line-up, but you can buy one with air conditioning from John Deere. They're not cheap, however. The John Deere 1585 TerrainCut will set you back $25,000, but at least you won't sweat in the sun.

46 Motorcycles Part I

The first motorcycle was invented by Gottlieb Daimler and Wilhelm Maybach in Germany in 1885. Since then there have been a wide variety of motorcycles, including steam-powered models and tandems with two separate seats. But who thought a caterpillar-track motorcycle was a good idea? The French, that's who.

The Mercier Moto Chenille, or motorised caterpillar, was designed by Swiss inventor Adrien Mercier in 1932. Mercier had a successful company building mopeds and motorbikes in France. His Moto Chenille had a tank tread in place of its front wheel. Powered by a 350cc engine, it reached a top speed of 40mph.

The French Army had been testing caterpillar-tracked motorcycles for a while, trying to improve upon its messenger motorcycles. Hoping Mercier's version had potential, it began testing the Moto Chenille in 1937.

Mercier's invention performed well on steep ascents, but its armoured windshield proved so cumbersome it had to be removed. The second test, not held until 1939, pitted the Motor Chenille against a Sevitame, the French Army's standard messenger bike. Tested on 90 miles of poor country roads, the Sevitame proved faster but the Moto Chenille did better off road. That is, until it had to ford a muddy trench, where it bogged down.

The French Army asked Mercier to make improvements. The Moto Chenille's low centre of gravity made it especially difficult to turn at slow speeds. However, Mercier refused to make changes, saying he'd already spent too much time and energy on the project. Not surprisingly, the military didn't order any, so only a handful of Mercier's inventions were built.

That didn't stop J. Lehaitre of Paris inventing his own caterpillar-track motorcycle. The 'tractor-cycle', as he called it, replaced both tyres with a single, continuous tank tread. This enabled it to travel over more varied terrain than a conventional motorcycle, but limited its top speed to 25mph.

The Moto Chenille designed by Swiss inventor Adrien Mercier in 1932. (Courtesy Haas Moto Museum)

Besides looking like it weighed a ton, the tractor-cycle was difficult to steer, as is often the case with this type of vehicle. When it was featured on the cover of *Modern Mechanix* in 1938, the magazine said it needed only a machine gun before it 'could be used by dispatch riders or military units to ride over shell-torn terrain'. The fact that Lehaitre's invention was never mass produced means it never got the chance.

The best use of a caterpillar-tracked motorcycle would seem to be in snow. That's what BMW thought in 1936 when it developed the Sneekrad, or snow motorbike. By far the best looking of all the early tracked motorcycles, it even came with a shiny, black sidecar. BMW used its R12 production bike as the foundation, but

Lehaitre's
tractor-cycle.

A BMW
Sneekrad.

once again steering turned out to be a problem – one reason the Sneekrad never got past the prototype stage.

The Germans and French weren't the only ones to build tractor-tread motorcycles. A Russian engineer named Eduard Luzyanin built one in 2020. Called the Khomyak, or Hamster, the all-terrain vehicle was assembled by cannibalising an engine from a Chinese motor scooter and a caterpillar track from a Buran snowmobile.

The Khomyak is only slightly larger than a pocket motorcycle. Short and squat, it looks uncomfortable to ride. It's also a monster to steer. Although it weighs only 187lb, weight distribution is your only option for steering. But the Hamster's caterpillar track is so wide, and the vehicle so close to the ground, it's hard to lean on. Plus, its recommended speed is only 12mph, meaning the only direction you're likely to go in is a straight line. That said, you can't buy a Khomyak. They're custom made and Luzyanin has only built two. This is just as well given its limitations.

A Mercier Moto Chenille can be seen at the Haas Moto Museum in Dallas, Texas.

47 Motorcycles Part II – Monocycles

Emergency room doctors often call motorcycles 'donor cycles' because of the frequency with which they kill their riders. It stands to reason that reducing a motorcycle's wheels to one would make them even more dangerous. That said, at least half a dozen inventors over a span of twenty years came up with their own version of a one-wheeled motorcycle, which many called the monocycle.

One of the first monocycles to be motorised (there were plenty of pedal-powered ones during the nineteenth century) was the

SUNDAY POST-DISPATCH MAGAZINE

ST. LOUIS, MO. JUNE 2, 1912

A MILE A MINUTE ON A UNICYCLE

REMARKABLE MACHINE INVENTED IN ST. LOUIS

RIGHT TO LEFT— CLINTON T COATES, INVENTOR OF UNICYCLE; WILLIAM McDONALD AND WM HUNTER.

The Coates Unicycle.

Garavaglia. Not much is known about this early entry other than it debuted in Italy in 1904.

A slightly more extensive paper trail exists for the Coates Unicycle, including a 1912 patent. Designed by Clinton T. Coates of St Louis, Missouri, its most unusual feature (besides sitting inside a huge, single wheel) was the giant push propeller mounted in the rear. Twin skids on either side kept it balanced, while a third stopped the propeller hitting the ground. Although there's no photographic evidence that Coates built and tested the Unicycle, correspondence suggests he did exactly that.

Alfred E. D'Harlingue lived in St Louis at the same time as Coates, but they appear to have been rivals rather than collaborators. D'Harlingue built his monocycle based on Coates's design with a few important improvements. Like Coates's Unicycle, the D'Harlingue Monocycle had three skids to keep it upright. A later version increased the number to five, three of which had wheels to reduce friction. Importantly, D'Harlingue mounted his twin-bladed propeller on the front of his invention rather than the back. Although the device had a steering wheel that changed the angle of its propeller, steering was largely accomplished by shifting one's weight from side to side just like riding a motorcycle. *Popular Mechanics* featured the D'Harlingue Monocycle on its cover in 1914. Said to reach a top speed of 67mph, it went through several iterations, but was never commercialised.

Alfred E. D'Harlingue and his monocycle.

Then there was the Christie Monowheel. Professor E.J. Christie, from Marion, Ohio, claimed his invention would 'reach a speed of at least 250 miles per hour, and possibly 400'. His Monowheel was actually five wheels, with only its centre wheel touching the ground. The operator sat on top of the main wheel's axle, which had two 500lb wheels on either side. These wheels were said to be gyroscopic and rotate at 90rpm to keep the vehicle balanced. When the driver wanted to turn, a steering wheel shifted the position of the two flywheels, enabling him to do so. Evidently, Professor Christie had trouble finding a tyre big enough for his Monowheel's main wheel, which is not surprising given it was 14ft high.

Professor Christie's Monowheel was built in Philadelphia, weighed 2,400lb and was powered by an aircraft engine generating 250hp, or so he claimed. The professor happily posed for photographs inside the contraption, which appeared on the cover of *Popular Science* in 1923. But despite his promises, Christie never demonstrated a working model – a pretty safe bet it didn't function as planned.

OOO

Davide Cislaghi* invented the Motoruota, Italian for motorwheel, in Milan in 1923. Described by various publications as a former electrician and motorcycle policeman, Cislaghi had a serious run spanning more than a decade. His Motoruota not only went through several versions, each one better than the last, but he also had some success commercialising his invention, although it never caught on in a big way.

Like the vast majority of monocycles, Cislaghi's Motoruota (he also called it a Velocita) was composed of two concentric rings: an outer pneumatic tyre and an inner ring that supported the driver's seat, engine and transmission. The inner ring rotated inside the larger ring on a small set of wheels. When the engine was engaged, the inner ring would climb up the inside of the tyre, propelling the vehicle forward in a manner that made it appear to glide magically across a surface.

The Motoruota was said to have a top speed of 100mph. It also had remarkable fuel economy, travelling 280 miles on a single

---

* His name has also been spelled Gislaghi.

gallon of petrol. This was an important benefit given its fuel tank was small. Additionally, the invention's low centre of gravity made it easy to turn. All the rider had to do was lean right or left. Cislaghi also said he could accommodate the different height of drivers by increasing or decreasing the size of the Motoruota's tyre.

The Motoruota, or one-wheel motorcycle, invented by Davide Cislaghi. (BNA Photographic/ Alamy Stock Photo)

There were doubters, of course. There always are where WETech inventions are concerned. But Cislaghi silenced them after riding his Motoruota at high speed from Milan to Rome.

Cislaghi demonstrated his invention in Italy, France and the UK, attracting lots of attention. At one point, he even planned on bringing the Motoruota to the United States to show its 'adaptability to American highways'. He also appears to have sold at least one to a Swiss engineer, who was photographed in 1931 riding it in Arles, France, while travelling to Spain. Although the Motoruota never caught on, it looks like a lot of fun.

ooo

If anybody needs proof that history repeats itself, consider the Dynasphere.

Its inventor, Dr John Archibald Purves, was a highly qualified Scottish engineer whose day job was supplying electricity to English communities. But electrifying the countryside didn't make Dr Purves famous; inventing the Dynasphere did.

Dr Purves claimed that by eliminating three out of four wheels, the Dynasphere, also called a monocycle, 'reduced locomotion to the simplest possible form', making it 'the high-speed vehicle of the future'.

Dr Purves built several prototypes, including a scaled-down prototype powered by an electric battery. There were at least two full-sized, petrol-powered versions, one with a two-handed device that acted like a throttle, the other with a steering wheel.

J.A. Purves's Dynasphere carrying eight passengers. (Colin Waters/ Alamy Stock Photo)

Both full-sized versions were open-air two-seaters, meaning passengers got a face full of rain, wind or dirt depending on driving conditions. There's a terrific photograph taken in 1932 of the inventor's son leaning out of a Dynasphere as he demonstrates how to make a turn. Bald, with two tufts of hair standing straight up on either side of his head, he looks Larry Fine of Three Stooges fame.

The Dynasphere didn't need a powerful engine to generate speed given the efficiency of its design. This could be a problem when it came to braking because the only way to stop the 1,000lb Dynasphere was to turn its engine off. Another obvious discomfort was that the driver's seat rocked back and forth when accelerating or slowing. This phenomenon made the driver look like a hamster in its exercise wheel immediately after stopping.

In a 1932 Pathé newsreel, a woman (possibly Dr Purves's wife or mother) sits in the Dynasphere explaining how it works. The invention clearly wobbles at the start despite its tread being wider than any previous monocycle, but gains stability as it increases speed.

Dr Purves was granted a US patent for his invention in 1935. That same year, a cover story in *Meccano* magazine waxed enthusiastically about the Dynasphere, saying it 'possesses so many advantages that we may eventually see gigantic wheels … running along our highways in as large numbers as motor cars do to-day'.

But this wasn't to be. Difficult to steer, the Dynasphere was also hard to brake. As a result, the invention remains a poster child for White Elephant Technology.

48 & 49 Hugo Gernsback: The Office Isolator and Teleyeglasses

Inventor, author, editor, prognosticator, publisher, crook; Hugo Gernsback was quite a character. Considered the founding father of twentieth-century science fiction, Hugo founded and edited *Amazing Stories*, the first magazine devoted to the genre.

Gernsback was fascinated by electricity as a child, experimenting whenever he got the chance. As a university student, Gernsback studied electrical engineering in Germany. There he perfected a portable radio-telegraph transmitter and high amperage, dry-cell battery that was the most powerful in the world. Convinced it would make him wealthy, Gernsback emigrated from Luxemburg to the United States in 1904 to market his invention. Twenty years old, he arrived with only $100 in his pocket.

After patenting his battery, Gernsback sold the rights to the Packard Motor Company. Using the proceeds, he started a

Hugo Gernsback with his Teleyeglasses. (Alfred Eisenstaedt/The LIFE Picture Collection/Shutterstock)

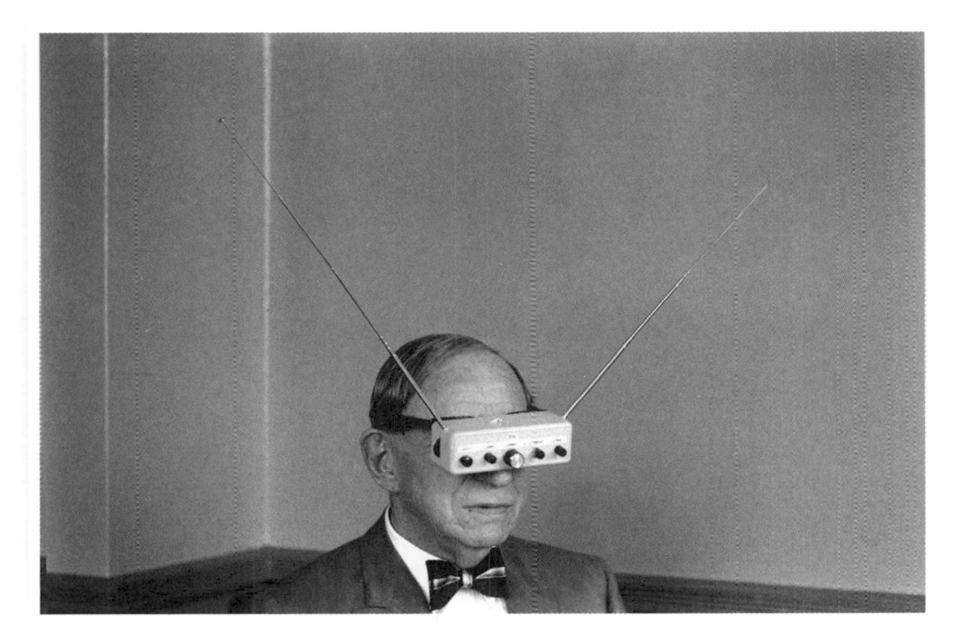

company to import electric supplies and radio components from Europe. He then assembled the parts and sold them as the world's first home radio set, creating a hobby market where none had existed. By 1908, the sixty-four-page catalogue he'd created to sell his products had evolved into *Modern Electrics*, the world's first magazine for amateur wireless operators.

Gernsback wasn't shy about his success. He affected a monocle, wore tailor-made suits, and was seen at all of New York's best theatres and restaurants. In 1924, he started several more magazines related to amateur radio, science and inventions. One of these put Professor E.J. Christie's Monowheel on its cover. Two years later, Gernsback started *Amazing Stories*, the world's first science fiction magazine.*

Billed as 'Extravagant Fiction Today ... Cold Fact Tomorrow', *Amazing Stories* became an amazing success. But Gernsback lost control of the magazine after losing a lawsuit. Undeterred, he immediately started two more science fiction magazines, despite it being the early days of the Depression.

Gernsback published more than fifty magazines during his career. The majority were hobbyist publications related to science, technology and invention. But the ones people remember today published stories devoted to science fiction. In keeping with his ego, he featured his name on the cover of his magazines while penning their editorials. He even put an illustration of himself watching television on the 1928 cover of *Radio News* well before the device became popular.

Gernsback had a reputation for paying his writers late or never. His sharp business practices led some of them to call him 'the Rat'. Nor was he an inspired writer. His two sci-fi novels are painful to read.

Gernsback had his hand in a lot of things. In addition to pioneering amateur radio in the United States, he co-founded a radio station in New York City. He was also a prolific if sometimes whimsical inventor. His inventions include a combination electric hair brush and comb, and the flame-throwing tank. But among Gernsback's most famous creations is the Office Isolator.

---

* Gernsback originally called the genre 'scientifiction'.

Amazing Stories. Note Gernsback's name in the upper right-hand corner.

The Isolator was a helmet you wore while working at your desk to avoid interruptions. As Gernsback wrote in the July 1925 edition of *Science and Invention*:

> Perhaps the most difficult thing that a human being is called upon to face is long, concentrated thinking. Whether you are a lawyer … pleading a special case, [or] an inventor with an intricate problem to be solved – assiduous concentration on the subject becomes necessary.

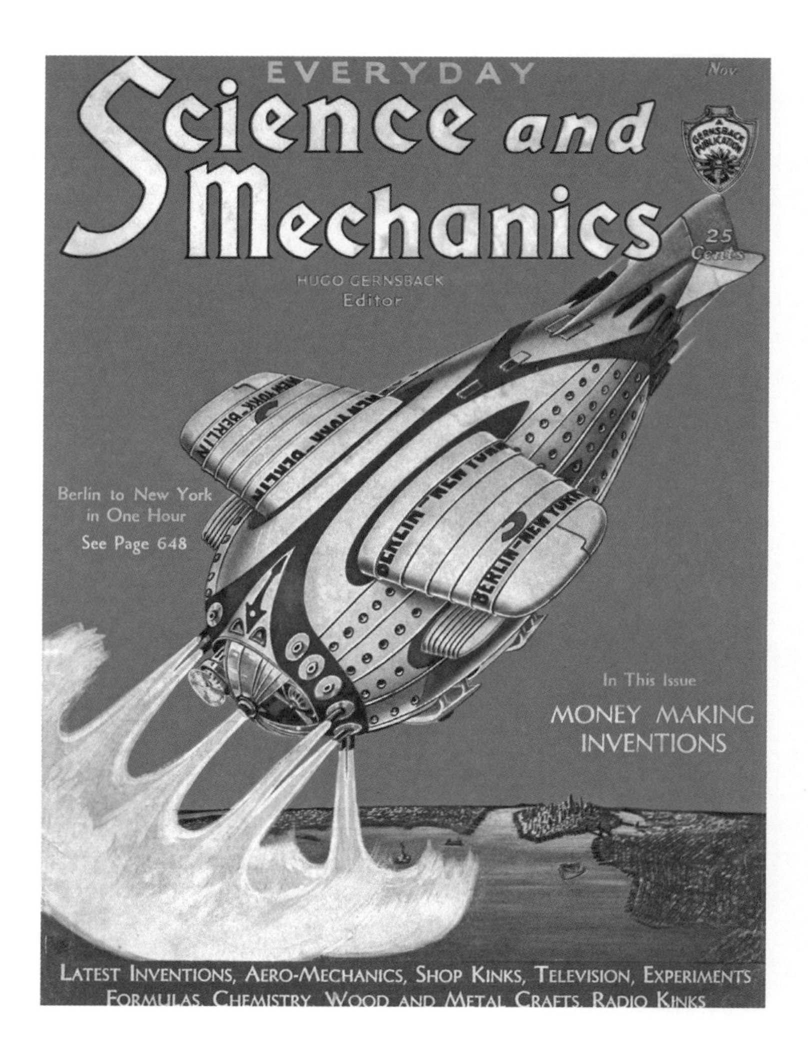

Science and Mechanics, November 1931.

The Isolator is a terrific example of an invention nobody asked for, especially since it looks like something a deep sea diver might wear. Made of wood, lined in cork and covered with felt, it had two tiny eyeholes made of glass and a tube connecting where the mouth should be to an oxygen tank.

Gernsback promised that, once outside noises were eliminated, 'The worker can concentrate with ease.' But he does not appear to have commercialised the Isolator since no patent can be found.

The Isolator wasn't Hugo's only odd invention. In 1936, he imagined a pocket-size portable TV. Unfortunately, the technology didn't yet exist to build a prototype. That didn't happen until

The Office Isolator.

1963, when Gernsback directed his employees to build a mock-up of what he called Teleyeglasses.*

Gernsback's Teleyeglasses looked like a cross between a View-Master and a virtual reality headset (see photo on page 187 for details). A photo in *Life* magazine shows him wearing the device sitting next to a bust of Nikola Tesla, another inventor who had difficulty later in life commercialising his inventions.

Gernsback's Teleyeglasses were a pocket-sized, battery-operated, portable TV with a separate screen for each eye. Weighing a little less than a third of a pound, the glasses came with miniature cathode-ray tubes and a pair of antennae

* Captions for photos on the internet often mistakenly call them Television Goggles

protruding from the top, making the wearer look like Ray Walston in *My Favorite Martian*. Because there was a separate screen for each eye, the device could display stereoscopic images much like today's 3D glasses.

Gernsback was 78 and more or less retired when the article appeared in *Life*. Although he said millions 'yearned for his invention', he never commercialised it.

Gernsback's early inventions made him plenty of money, but he was more famous for his predictions. In addition to predicting radar, air-conditioned homes and offices, and microfilm, he predicted TV would become an 'everyday affair', including televised baseball games. Then again, he also thought cemeteries could be eliminated by launching the deceased into space.

Although many scientists were inspired as children by Gernsback's publications, he was never fully accepted by the scientific community because, well, many people thought he was nutty. His theatrical behaviour and constant future-fying didn't improve his credibility. Still, Marconi, Edison, Tesla and Goddard corresponded with him, in part because of the free publicity Gernsback's publications could offer them.*

Gernsback held eighty patents over the course of his life, but despite many accomplishments, he was not easy to live with, which explains why he was married three times. Hugo passed away in New York City in 1967 aged 83. In his honour, the annual literary awards for the best science fiction are called the Hugos. He may be the father of science fiction rather than science fact, but he died after making a bundle from nothing more than his imagination.

* Gernsback was such a fan of Tesla that he commissioned a death mask when the inventor died, publishing photos of it in his magazines.

Chapter Nine

The Elephants' Graveyard

> We are all failures – at least the best of us are.
>
> J.M. Barrie

Many WETech inventions come to an ignominious end. Some disappear without a trace, while others devoid of dignity lie rusting in a scrapyard. Occasionally, one is rediscovered, rescued and restored but more often than not they're put in a museum to be laughed at rather than respected.

So where do WETech inventions go to die?

LeTourneau's first two overland trains (the progenitors for the Sno-Freighter and Sno-Train) were the VC-12 Tournatrain and the TC-264, also called the Sno-Buggy. Sadly, their fate remains unknown. As for the VC-22 Sno-Freighter, its locomotive cab and several trailers sit abandoned on the side of an Alaskan road. The Sno-Freighter's final resting place is not far from the salvage yard where the Sno-Train once mouldered. It's a sad end to a remarkable vehicle, but you can still see it when driving on the Steese Highway south of Fox, Alaska.

Not all of LeTourneau's overland trains outlived their usefulness. Bob Chandler found four massive, 10ft-high tyres from one of the overland trains in a Seattle scrapyard. After fitting them to a pick-up called *Bigfoot*, he secured the title of world's tallest monster truck, according to Guinness World Records.

An abandoned Sno-Freighter.

Japan's underwater aircraft carrier, the *I-401*, was captured by the US Navy after the Second World War. Sailed from Japan to Pearl Harbor, it was studied to determine whether any of its technology should be incorporated in the navy's next-generation subs. After a thorough examination, it was then unceremoniously torpedoed off Barbers Point.

The *I-401* lay undiscovered on the sandy sea bed for nearly sixty years. Then, in 2005, a University of Hawaii submersible named *Pisces IV* located and photographed the sub 2,600ft underwater. It lies there still, slowly rusting away.

The US Navy's flying aircraft carrier, the USS *Macon* (ZRS-5), crashed after suffering a catastrophic structural failure somewhere off the California coast in 1935. Missing for fifty-five years, it wasn't discovered until 1990, when the US Navy used the *Sea Cliff*, a deep-diving submersible, to locate the wreck site. By then, most of its giant airframe had disappeared. All that was left were eight Maybach engines, each as tall as a man, a few aluminium chairs and four Sparrowhawk biplanes sitting upright and intact on the ocean floor, where they remain today.

An abandoned Soviet jet-powered SVL train.

The Soviets' jet-powered train, the SVL, sat on a railway siding for many years. When the Tver Carriage Works erected a monument in 2008 celebrating its 110th anniversary, it featured a replica of the jet train's nose and twin engines. But the actual SVL high-speed railcar sits behind the Kalinin rail factory where, like the USS *Macon* and *I-401*, it is quietly decaying.

A triple-mount anti-aircraft gun on the wreck of the *I-401*. (Hawaii Undersea Research Laboratory Archive)

The sail of the sunken *I-401*, showing an open hatch leading down into the sub. (Hawaii Undersea Research Laboratory Archive)

Two of the USS *Macon*'s (ZRS-5) aircraft resting on the ocean bottom off the coast of California. (MBARI-NOAA/MBNMS-Ken Israel Integral Consulting Inc.)

50 The Caspian Sea Monster

Finally, we come to Russia's one of a kind, the Lun-class ekranoplan, a giant aeroplane classified as a ship by the International Maritime Organisation.

Intended to deploy troops quickly in an amphibious assault against NATO forces, the ekranoplan used the 'ground effect principle' to skim across the ocean on a cushion of air at a height no greater than 15ft. Truly enormous, it was the largest aircraft in the world when built, and remains the second largest today. With eight turbofan engines mounted on two forward canards, it achieved a speed of 340mph, despite weighing 380 tons. It also carried six anti-ship missiles in launch tubes on top of the hull to defend against attack.

The abandoned Caspian Sea Monster. (Oleg Znamenskiy/ Alamy Stock Photo)

The Soviet ekranoplan was nicknamed the 'Caspian Sea Monster' by US intelligence shortly after it entered service in 1987. However, its usefulness was limited. Contrary to popular belief, it was not invisible to radar. It also provided a very bumpy ride when seas were rough. A second prototype intended for rescue missions was never completed before the collapse of the Soviet Union, which led to the programme's cancellation. Left to decay at the Kaspiysk naval base for thirty years, it was finally rescued by three tugs, which towed it on a fourteen-hour journey to the western shore of the Caspian Sea. Intended as a tourist attraction for a military museum yet to be built, it has been sitting on the beach near Derbent for more than a year. Money is tight according to the local mayor, so keep your fingers crossed that the Caspian Sea Monster will one day be restored.

Chapter Ten

Just Around the Corner: The Future of WETech

> There are some defeats more triumphant than victories.
>
> Michel de Montaigne

The media often exaggerate how close a prototype is to being commercialised. In some cases, an invention is too far ahead of its time to attract interest. But for many, there's not a strong economic case, enough consumer demand, or a clear regulatory path to approval to get past the prototype stage. Having said this, there are a surprising number of fully functioning flying cars and jet packs lining up to hit the market. And flying cars aren't the only things we've been promised that have made great strides. Robots are already de rigueur in manufacturing. They may not be the Republic robot we were promised in movie serials during the 1940s and '50s, but as algorithms, artificial intelligence and machine learning advance, we can expect to see more of them in the future. That doesn't mean we're going to have Rosie from *The Jetsons* cleaning up after us anytime soon, or for that matter, fully autonomous cars. But as advances continue, it's reasonable to expect WETech inventions will follow in their path.

Not everything we've been promised has achieved widespread acceptance. Monorails were supposed to be a regular part of daily life, but save for a few examples in Germany, the United States and Japan, they never took root in any appreciable number. The same holds true for hovercraft cars. Although the technology has been successfully adapted for seagoing ferries, there aren't that many of them. Wing suits have also been perfected. They won't stop you from hitting the ground as Franz Reichelt hoped. They will make your descent more fun until you pop your parachute, but they're not for everybody. Still, there's no upward limit to how many WETech inventions the future will bring. As you read this, thousands of inventors labour in obscurity developing what they hope will become the next big thing.

Arthur Radebaugh, a commercial illustrator who illustrated the syndicated comic *Closer Than We Think!*, knew this better than most. The influential Sunday newspaper strip, which showcased future inventions, reached nearly 20 million people between 1958 and 1962. Radebaugh's predictions were sometimes on the mark (robot warehouses, computer navigation and electronic Christmas cards), sometimes wildly off (atomic cars, jet-pack postal delivery and mining on the Moon), but they were always inspirational. An excellent example of twentieth-century techno-optimism, they were another way of viewing White Elephant Technology.

Radebaugh's vision, like a lot of WETech inventions, was ahead of its time. Eventually superseded by *Tomorrowland*, *The Jetsons* and *Blade Runner*, *Closer Than We Think!* was forgotten when Radebaugh died in 1974. But that doesn't mean he wasn't on to something.

Let Us Now Praise Infamous Men

It takes more than imagination to invent something new; it takes courage. Nobody knows which inventions will succeed. For every success we read about, there's a scrapyard full of failures of which we never hear.

But failure is a necessary part of success. Without it progress would be stymied. Sir Ernest Shackleton never succeeded in reaching the South Pole, but his failure turned into one of the most celebrated achievements in polar history.

An invention not only contains an inventor's hopes, dreams, desires and ambition, it's a stepping stone to a better life. Still, their sacrifice and hard work rarely go recognised. This is why they must be celebrated. We need their failures to pave the way for success, otherwise nothing will be ... closer than you think! Fortunately, they'll always be with us. Just like failure.

About the Authors

John J. Geoghegan

John J. Geoghegan specialises in reporting on unusual inventions that fail in the marketplace despite their innovative nature. His articles on WETech inventions have appeared in the *New York Times* Science section, *Popular Science* and the Smithsonian's *Air & Space* magazine, among other publications. He is also the author of two previous books about White Elephant Technology: *Operation Storm: Japan's Top Secret Submarines and Its Plan to Change the Course of World War II* (Crown, 2013), and *When Giants Ruled the Sky: The Brief Reign and Tragic Demise of the American Rigid Airship* (The History Press, 2021). John currently serves as Director of the Archival Division at the SILOE Research Institute in Marin County, California.

Eric Miles

Eric Miles is a researcher and writer working at the intersection of photography, history and the printed page. He is currently the Visuals Editor for *Vanity Fair*. He lives in Brooklyn, New York, with his son and two cats.

Bibliography

Books

Geoghegan, John J., *Operation Storm: Japan's Top Secret Submarines and Its Plan to Change the Course of World War II*, Crown Publishers, New York, NY, 2013.

Geoghegan, John J., *When Giants Ruled the Sky: The Brief Reign and Tragic Demise of the American Rigid Airship*, The History Press, Gloucestershire, 2021.

Knabenshue, A. Roy, *Chauffeur of the Skies*, unpublished manuscript, National Air and Space Museum, A. Roy Knabenshue Collection, Washington DC.

Mecklem, L. Guy, *The Autobiography of Llewellyn Guy Mecklem*, unpublished manuscript, Seattle Public Library, Seattle, WA.

Reid, Bruce D., *The Flying Submarine: The Story of the Invention of the Reid Flying Submarine, RFS-1*, Heritage Books, Inc., Westminster, MD, 2004.

Santos-Dumont, Alberto, *My Air-Ships: The Story of My Life*, Kessinger Publishing, LLC, Whitefish, MT, 1904, reprinted 2010.

Articles

Geoghegan, John J., 'Row, Row, Row Your Airship', Smithsonian *Air & Space* magazine, June/July 2014, Vol. 29, No. 2, p.18.

Geoghegan, John J., 'Full Steam Ahead', *American History* magazine, April 2017, Vol. 52, No. 1, pp.26–33.

Geoghegan, John J., 'Man-Carrying Kite', *Aviation History* magazine, September 2013, Vol. 24, No. 1, pp.14–15.

Geoghegan, John J., 'The Dream of Steam', *Aviation History* magazine, January 2017, Vol. 27, No. 3, pp.36–41.

Geoghegan, John J., 'Boat, Moved Only by Waves, Sails to a Seafaring First', *The New York Times*, 8 July 2008, Science section, p.D2.

Geoghegan, John J., 'Designers Set Sail, Turning to Wind to Help Power Cargo Ships', *The New York Times*, 28 August 2012, Science section, p.D4.

Geoghegan, John J., 'Long Ocean Voyage Set for Vessel That Runs on Wave Power', *The New York Times*, 11 March 2008, Science section, p.D3.

Geoghegan, John J., 'Wave Runner: A New Propulsion System for Boats Ditches the Diesel', *Popular Science*, March 2008, Vol. 272, No. 3, p.35.

Want to Recommend a WETech Invention?

If you know about a WETech invention, or are an inventor interested in promoting an especially unusual or innovative invention, please submit it for consideration to our WETech website at www.WhiteElephantTechnology.com, or email us the details at WhiteElephantTechnology@gmail.com.

Who knows? It might just appear on our website or in our next book.

By the Same Author

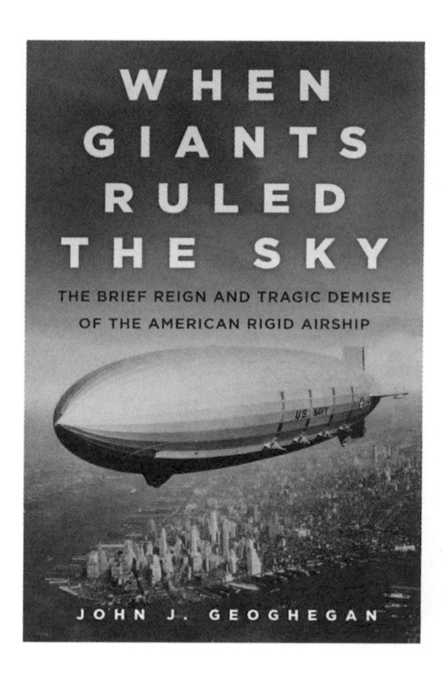

978 0 7509 8783 7

When Giants Ruled the Sky is the story of how the American rigid airship came within a hair's breadth of dominating long-distance transportation. It is also the story of four men whose courage and determination kept the programme going despite the obstacles thrown in their way – until the US Navy deliberately ignored a fatal design flaw and brought it crashing back to earth.